MOONRUSH

MOONRUSH

Improving Life on Earth with the Moon's Resources

by

Dennis Wingo

An Apogee Books Publication

All rights reserved under article two of the Berne Copyright Convention (1971).
We acknowledge the financial support of the Government of Canada through the Book Publishing Industry Development Program for our publishing activities.

Published by Apogee Books an imprint of Collector's Guide Publishing Inc., Box 62034, Burlington, Ontario, Canada, L7R 4K2, http://www.cgpublishing.com

Printed and bound in Canada

Moonrush - Improving Life On Earth with the Moon's Resources by Dennis Wingo

ISBN 1-894959-10-8 - ISSN 1496-6921

©2004 Dennis Wingo

Contents

Dedication

This book is dedicated to my fellow voyager toward the asteroids, Bob Asprey who has helped provide the support over the years to make this book possible.

To my Nikki love. Thank you for bringing your sunshine into my life!!

Acknowledgements

Many people helped to inspire this book and helped to get me to this point in my life. I want to thank you all, just a few of which are listed here.

An extra special thanks goes to Mark Maxwell whose unique talents provided visualizations on which to hang my ideas.

Gordon Woodcock, space is in his blood
Rich and Rob Godwin, brothers in arms
Dr. Charles Lundquist, his slide rule took us to space
Dave Christensen, who knows the history, he helped make it
Dr. Gene Scott, who taught me about faith and commitment
Dr. David Webb, who knew that limits were all in the mind
Dr. John Lewis, the best asteroid guy around and one of the inspirations for this book
Dr. John Gilbert, who knew that concrete canoes and space went together
Brian Kelly, he understands what it takes to make this work
Lewis Peach, Charlie Gunn, Karen Poniatowski, and George Levin, took pity on us
Cheryl Alexander, all that work was worth it
Walt Anderson, a true believer who put his money where his mouth is
Nick Esser, the best boss ever in the computer industry
Andy Reichart, one of the unsung soldiers in the computer age
Vector Graphic, Inc., gone but not forgotten, we had a lot of fun
Dr. Bob and Lore Harp, although parted, together helped to change a world
Shubber Ali, thanks for the help and comments
Libby Workman, she knew before we did
Milton H. Thomas, a great teacher who knew when to look the other way
Margurite Gann, my eighth grade science teacher who wanted me to do better
Steve Jobs, thanks for the laptop and the best computer ever
Al Smith, all around great guy who understands where I am going
Shatas, wrote the first check and then another and then another
Breezely Bruhns, who dodged the birthday cake and helped make this possible
Ken Kledzick, twenty eight miles to Nathan's is a standard measure
Pete Worden, who knows how to make this work
Nathan Hershfeld, who knew that twenty eight miles was not enough
Jay Hardin, I never knew a car could do that until he did it
Don Heptinstall, a friend who did not make it out of the 70s
James Dennis Bell, my pal gone to his reward
David Smith, forever young
Brother Ryland, he believed and helped me to as well
Graysville Alabama, a lot of stories and my hometown
My mother, my brother, and all the rest......

Prologue

Recently, the World Wildlife Federation declared that it would take the equivalent of two more Earth's to sustain our planetary population at the level of affluence that the western world enjoys. Today we live in a world of six billion people who are gobbling up our planet's resources at a tremendous and accelerating rate. The advent of cheap energy in the form of oil has been the key factor that has enabled us to develop a planetary civilization of unprecedented size, complexity, and comfort. However, that same energy is accused of altering our climate and at best will be depleted within a hundred years. Additionally, tremendous amounts of water and air pollution are generated by the extraction of increasingly minute amounts of nickel, copper, aluminum, and other primary metals from the Earth. In other areas, resources are strained; from the fisheries of the North Atlantic to clean water in India and China. Indeed, many in the environmental movement believe that we have gone beyond the limits to growth and that it is only a matter of time before the whole system collapses.

"More Worlds" is what this book is about. While in this solar system there are no more Earths, there are several planets, hundreds of Moons including our own, and millions of smaller planetoids that can provide resources for the betterment of life here on the Earth. This book will concentrate on the economic development of the world that is closest to us in space: our own Moon. I will outline a scenario about how the resources of the Moon can dramatically increase our planetary wealth, to help transcend our dependence upon oil, provide for a diversified energy and resource future, and provide the means to improve all of our lives. The technologies and resources developed there can also make the grand human voyage to Mars much more than what we were given in Apollo—flags and footprints.

This scenario is intended to broaden the participation in space efforts beyond the solely scientific approach that is the hallmark of NASA. NASA will be a vital contributor of space-specific technology and will be a valuable participant in the enterprise, but if we are going to actually develop these resources as an economic engine, the effort must include, to the maximum extent possible, the participation of private enterprise and more than a few government employee-scientists-explorers. The eventual goal is for the economic development of lunar resources to contribute taxes to the treasury and to help tilt the balance of payments (the ratio between imports and exports) to one more favorable to the United States. If we were able to fully develop technologies associated with fuel cells (Lunar Platinum Group Metals) and the "Hydrogen Economy", we would be able to use the vast resources of methane ice located at the edges of the North American continental shelf. Dropping our dependence on foreign oil would eliminate the dramatic deficit today between imports and exports. Recent advances in technology make this much more than just a dream. Indeed the Moon and Mars could become the testing ground for the full implementation of the Hydrogen Economy.

In the past 30 years since the end of the Apollo Lunar missions, a technological revolution has taken place that has given us satellite television and radio, and a personal computer in almost every home in the developed world, connected to a global Internet whose impact is still growing in our lives. A profound digital divide has developed between Silicon Valley and the aerospace community to the detriment of aerospace. A simple example is that the code that operates our desktop computers is orders of magnitude more complex than that used in computers on spacecraft. This divide will be examined and examples will be given of how the dramatic advances in the world of silicon devices and the skills of Silicon Valley can help lower the costs of space hardware and enable The Second Space Age. It is even very possible that the first landing on Mars will come from a space vehicle that is built on the Moon. With the advances in tele-presence, computer controlled fabrication, and human participation this may be the most cost effective way to open Mars for human exploration and development.

Lowering the costs of executing this vision of space for the Moon and Mars is absolutely necessary and we must look beyond the traditional NASA/contractor model to do this. In the past the U.S. government has provided incentives for entirely new modes of transportation. In the early days of the U.S. as a nation, canals were built to speed the transport of goods across the northeast. In 1804 Robert Fulton's steamship was given statutory support from the state of New York that enabled private risk capital to bring the steam age to shipping. The railroads were similarly enabled by government policy in the Railroad Act of 1862 to bridge the North American continent with bands of steel.

Early in the 20th century, Teddy Roosevelt's administration and congress passed laws that enabled the construction of the Panama Canal bringing the U.S. to commercial parity with the great nations of Europe. Succeeding administrations created similar incentives and passed laws to enable the rise of the U.S. aerospace industry that has helped make the U.S. the world's greatest superpower. As the 21st century dawns, we must examine these historical precedents and implement similar ones that do not bankrupt the treasury and enable private enterprise to enter this new domain. As the 20th century was the Century of Flight, the 21st century should be the Century of Space. This is a proper role for government: to foster, facilitate, and provide incentives to enable private enterprise to open up a new world for development. This is a role that transcends NASA's solely scientific efforts although NASA will be a vital part of this process.

There are many who would say that today is not the time to go to the Moon or on to Mars. It has been said since the end of the Apollo program that our national treasury would be better spent on education, or healthcare, or the environment. This argument did not sway the Congress or Lincoln in the depths of the War between the States when, in the midst of fighting for the life of the nation, money was spent and laws were passed for the completion of a "National Railroad," to bridge the North

American continent. When the very future of the nation was in doubt, and thousands were dying per day on the battlefields of the divided nation, these leaders looked a hundred years in the future and provided scarce funds to enable a better day for their posterity.

For a nation to provide for its citizens, it must create wealth. Education, healthcare, and the environment are all noble areas to spend taxpayer money, but without new sources of wealth, very few of those noble areas can be addressed successfully. On the Moon, Mars and the other bodies of the solar system there is wealth to help power our civilization for hundreds of thousands of years. This is our task today to provide for our posterity.

This is why we need to go to the Moon and on to Mars and do it now: to make life better for all of us on the Earth, not just for today, and not just for a hundred years. The World Wildlife Federation was right; it does take more than one Earth to enable a prosperous future for all the people of the Earth. Fortunately there are literally millions of worlds just in our solar system for our use. This can be the best legacy that our generation leaves the world: a way beyond the limits to growth, and toward a peaceful and prosperous future.

*   *   *   *   *

Chapter 1:

Why Space? Why the Moon? Why Now?

Defining the Issue

This book begins with a question: the question, "Why?" In writing about something such as the exploration and development of space and a return to the Moon, "why" is the most important question. Without answering, "why," inevitably *how*, *who*, and *when* are irrelevant. The first chapters of this book address "why," and in a slightly different manner than those who have come before. Typically, NASA begins with science, designs a space transportation architecture, then a base, and possibly, talks about resource development. In this book we begin with commercial activity, settlement by humans, accommodate the government, and provide data to the scientists. To do this we have to start at why and connect it to our lives today and see how commercial development and human settlement of the Moon can help to solve critical Earthly problems, especially resource depletion and possible global climate change.

"Why" must be prefaced with a reason to ask the questions above in the first place. The reason that I postulate here is:

> *The exploration and development of space, including a return to the Moon and on to Mars, must bring concrete benefit to the people of the United States and the world, to transcend the problems of today, and improve our lives for ourselves and our posterity.*

This benefit must go beyond science or the merely intangible. It must be something that we all can see and point to each and every day. It must also go beyond what space has done for us in the past in order to justify the expense in national treasure that such an endeavor requires.

Those of us who advocate the return to the Moon must justify our claim on the national treasury, and our desire for incentive laws to promote commercial enterprise, as others do for the things that they advocate. University level political science classes speak of the way in which *competing* interests of advocacy groups influence policy and work to get laws passed and funds appropriated for various activities deemed useful to those groups and the people of the United States. The huge number of competing interests and their often conflicting goals results in a diffusion of power as these groups contend with one another. This diffusion of power was intended by the founding fathers and is built into the structure of U.S. government.

The founding fathers did not trust the motives of those who seek to influence government. They began with the premise that people cannot be trusted to be good and will seek their own interests. From the separation of powers of the legislative, executive, and judicial branches of the government, to the influence of an individual person writing a letter, we are all advocates for or against something in government. This is the power of representative government and the Constitution; that we can, and do, exert these influences. However, to accomplish this, competing interests must ally with others to accomplish their goals, influence policy, and achieve the critical mass necessary to compel our representatives in government to pass laws that support these interests.

Today these competing interests are called "special interests." There are literally thousands of special interest groups in the nation and all seek to promote the interests of their members. These include unions, corporate lobbyists, farmers, senior advocates, environmentalists, as well as space advocacy groups. All of these groups fit the definition of special interest groups and they are all focused on their own agendas, which is right and proper for them. The term "special interests" is thrown about as a derogatory term, but it pretty much defines all the groups that are active in influencing politics, the good, the bad, and the ugly.

Our political system was designed to be inefficient at getting things done. The founding fathers thought that this would be the best way to delay the corruption of our form of government. So, in order for a new idea to reach the critical mass necessary to guarantee that the government takes serious action, it must be able to appeal to a broad cross section of society. In other words, people will support an idea if they believe in the utility of the idea and it provides clear value. This support translates into political power, and political power translates into action by the government.

This happens every day in the debate over national priorities. National defense, while much debated, is acknowledged to be necessary and is funded to the tune of several hundred billion dollars a year. The same is true of our health and welfare expenditures. There is broad agreement that these activities are valuable and thus the government acts to support these priorities. However, the more specific the interest, the less it appeals to the majority of the people. Space advocates since the Apollo era have been singularly ineffective in this advocacy activity. The reason for this is simple. The dreams of most space advocates reach too narrow a cross-section of the electorate. Space exploration is supported, but not in a way that translates into political action for a higher level of activity than what we have today. Examples of narrow interests reaching a wide level of support are common in the history of the U.S. and other countries.

In the early history of the U.S. and in Britain, canals were built; paid for by the government in order to promote commerce. A few years later, it was support from the government of the state of New York that enabled Robert Fulton to raise the capital to build the world's first practical steamship in 1807. Even during the desperate times of War Between the States, the "National Railroad," of the 1860s, was strongly supported financially by the government. This railroad provided clear value to the people in cutting the travel time across the country from months to a few days. In the 20th century, the government supported and paid for construction of the Panama Canal. More recently support, subsidies, and direct funding was provided to the fledgling air transportation industry and for the construction of the world's largest interstate highway system immeasurably improving the economic well being of the nation.

The unifying thread of all of these projects is the promotion of commerce and the solutions to national problems; something the founding fathers were intensely interested in promoting as they saw the positive effects of this support in the explosive growth of English industry. Not only did these government-supported activities promote commerce, they also promoted national unity, the safeguard of lives, and built the wealth of the nation. Space and a return to the Moon must fit within that same framework of broad national benefit if we are going to be successful in gaining the support that is needed to make it happen.

In the first space age (the age of the Apollo program), there was no clear link to the overall national interest beyond cold war propaganda. Even at the height of the Apollo program, the majority of scientists and industry did not support the NASA lunar exploration program, but thought that they could use the money better on their own research and development activities.[i] The American Association for the Advancement of Science magazine, *Science* Editor at the time, Dr. Philip Abelson, director of Carnegie Institution of Washington was at the forefront of this lack of support. He wrote in Science magazine that a poll of their membership came out 110 to 3 against manned missions going to the Moon. In an interesting statement related to this, their justification for this opposition was:

"We already know that there will be no objects of economic value to be brought back from the Moon or any of the planets"[ii]

This lack of support in the wider scientific and industrial community would be telling in the years after Apollo.

There was no broad support across the country for the potential of the lunar program to promote commerce. Wernher von Braun and Walt Disney produced general interest programming for the Disney audience in the 50s to garner public support, and others discussed this extensively, and yet funding did not appear. The original Apollo era space program (as opposed to what we have now) was driven by an urgent national priority to demonstrate the ability of our technology and political system in contrast to our Soviet competitor. In those terms, the Apollo program did exactly what it was supposed to do and was shut down after accomplishing its mission. In contrast to what I and other space advocates think, the majority of people today do not see space, its exploration, development, and settlement as very important to their future. They think space is cool and the pictures are nice, but it does not connect to them in their daily lives. However, it should, because space, the return to the Moon and beyond is at the core of our future prosperity and possibly our civilization's survival.

Making the Connection

Reconnecting space and the return to the Moon with our daily lives is what this book is all about. Space and a return to the Moon need to be reevaluated according to their potential to address problems that are broad enough to concern everyone. If returning to the Moon and on to Mars cannot justify this, then this exploration and development effort will not happen and deservedly so. The political support that is here today from the president and NASA administrator may go away at some point in the future and any public policy about space must be able to survive the political process long enough to be accomplished.

Therefore, it is instructive to look around to see what problems affect the broad spectrum of people here in the U.S. and around the world. What are the top five problems that affect the world today, and can space address them?

    Terrorism
    Global climate change
    Energy
    Pollution
    Overpopulation

These are in no particular order since many would dispute the order, and probably the list too, but this is a pretty good representation of the biggies. What in the world can space do for any one of these to make the problems materially better for everyone living here today as well as for the future? The central argument of this book is that space exploration and a return to the Moon can affect each and every one of the above problems much more efficiently than existing, or future, solely earthbound solutions. The five above are intrinsically intertwined and are governed by problems related to the limits of our Earth to carry our large population of humans. I will start the process of developing the arguments for the return to the Moon by focusing on one of the problems above and then looking at the effect that it has on the other four. That problem is energy.

Energy is the Key to the Future

Today energy is the key to our future. In the list of problems, energy, and the politics surrounding it, affect all of the other problems in either a direct or indirect manner.

Our use of fossil fuels impacts global climate. We don't know how much yet, but research by the climactic research community, especially paleoclimatologists, have conclusively shown that our climate has changed dramatically in times past and these past changes indicate that equally large changes will come in the future whether or not we use oil or whether or not humans even exist. So we as a global society have to be able to figure out how to deal with climate change while at the same time having availability of low cost, plentiful energy.

Energy goes to the heart of terrorism as well. If there were no oil in the Middle East, that problem would be considerably tempered. We as a country, and as a world, need a way out of the need to send troops to defend the very lubricant of our whole civilization. If it were possible, would it not be better to get our energy needs from space, or from space derived materials and terrestrially derived improvements based upon advances in space technology? This would take political pressure off of world leaders and would help create a more peaceful world which is obviously a desirable goal.

Energy is also central to population growth and pollution related issues. In some areas of the world, we still have food heated by dung fires, no electricity, and poor infrastructure easily overwhelmed by disaster. It is in these under-developed poor nations where population growth is highest. Our global society has helped to cut the death rate in these poor countries but the birth rate has not fallen. This imbalance is typically driven by poverty, lack of education, and economic opportunity. In contrast, there is a curious fact about wealthy nations. They make fewer babies and their population growth slows to replacement level. It is historically clear that with low cost energy comes wealth. The United States still uses almost a quarter of the fossil fuel supply on the planet, and while energy does not directly translate into wealth, what we do with energy here in the U.S. has helped to make our nation very wealthy. Would it not make much more sense to live in a world where everyone is prosperous, and by their own decisions, not coerced by governments, decided to have fewer children?

Current movements in the area of population growth use phrases like, "unprecedented cooperation and change in thought patterns are necessary to address population growth." This statement, along with others like it, are espoused by those who see no possible solution to our problems other than a dramatic retrenchment of the progress achieved by humankind over the past few centuries. Indeed, some of the most quoted writers make statements like "China cannot be allowed to achieve the same level of material prosperity that the U.S. enjoys." This logic is rooted in the supposition that we only have the resources of this earth and that China's demands on them will far exceed the available supply.

Space based resources can help to make their arguments moot while providing for the common good of the people of the Earth. This is why the return to the Moon is important, if for no other reason than that China has the same rights that we do to provide a prosperous future for its people, and we have no right to stand in their way. If the return to the Moon, the development of its resources, by private enterprise supported by the government, provides the way for this to happen, then we stand a good chance of avoiding the wars that are otherwise an inevitable consequence of a fight for the Earth's limited resources.

Where We Go From Here

If it can be shown that the resources of the Moon and near Earth free space can provide a "third way" beyond the current "what me worry," and the environmental doom-and-gloomers, then it behooves us to commit the resources to investigate this path. The issues that confront our global civilization are many, and those who study these issues either despair that we will ever find solutions and are waiting for the house of cards to fall, or suggest solutions that are incompatible with the continuance of individual liberty. There is a song from the rock band The Police that sums up these feelings, *"When the world is running down/ you take the best of what is still around."* Is this the kind of world that we want to give future generations? In the following chapters, some of the "what me worry" and "doom-and-gloom" arguments will be laid out and contrasted with what the future can hold with a space based economy. It is startling to look at some of the proposed solutions that form the mainstream of political thought today, and the unwillingness to look at alternatives that do not have severe long-term consequences. It is our duty as space advocates to present positive alternatives that we have great confidence will work. These alternatives will enable us to protect and improve the environment, give everyone on the planet the chance to have a high standard of living, and also to preserve, and even extend, individual liberty.

There is a vision of the future to be seen here. What the resources and location of the Moon can do to improve our daily lives is tremendous. The underlying premise is that we can use the Platinum Group Metals (PGM's) as the key component of fuel cells for cars and other power generation applications that will allow us to switch from heavy oil hydrocarbons to light hydrocarbons like methane ice, which is plentiful at the bottom of the continental shelf of the United States. This would save billions of dollars while allowing for a complete switch to the environmentally friendly hydrogen economy in a more controlled fashion when nuclear fusion's promise of nearly limitless energy is finally fulfilled.

Another huge benefit would be the use of the Moon's environment for industrial processes that require a vacuum. Today, industries as diverse as nanotechnology and metallurgy use vacuum manufacturing to make higher purity alloys and improve quality and operation of nano-fabricated components. Another possible use of this vacuum could be to advance the state of the art for engineering a working fusion reactor. Obtaining a vacuum on the earth is expensive, and today, a fusion reactor on the Earth has to be set inside of a vacuum chamber at great initial capital and operating expense. The natural vacuum on the Moon is a thousand times better than the best vacuum obtained on the earth.

With a lunar manufacturing facility, for every ton of oxygen produced, over 3 tons of iron is produced (from the release of oxygen from an iron oxide resulting in pure iron left behind) and depending on the heat applied, magnesium, silicon, aluminum, and titanium for use in manufacturing and construction. With this great surplus of metal, new alloys can be investigated and applied to the problems of spaceflight, as well as imported to the Earth. These advanced alloys can then be made into lightweight, powerful spacecraft that can then open the doors to the solar system as Mr. Bush suggested in his speech in January of 2004. Low cost access to space has been an Albatross around the neck of commercial space companies for over 30 years now with little hope of a solution in the near future. If launched from the Moon, most of the entrants for the X-prize could make it into lunar orbit. We need to suspend the doubting part of our brains long enough to think about the possibilities created by development of the Moon, rather than automatically assuming that nothing is there and that it is impossible to do anyway.

Many writers have grave doubts about industry on the Moon, but most of these writers are unacquainted with modern production methods and trends in the application of advanced materials. It is no longer a question of, "Can we do this?" It is, "Will we do this?" Space advocates need to understand that a narrow focus on science and exploration will not work and should work to understand how the Moon and its resources can contribute to the solutions to global problems. Without the Moon having these wide applications, it would not be worth the capital and change in laws that we seek. Fortunately it does, and in abundance.

For those who say that it will never be cost effective to bring payloads back from the Moon or indeed manufacture there, the question is, "Why not?" Intrinsically, the availability of inexpensive energy and efficient and lightweight solar power systems can easily drive a lunar economy, just as it does here on the earth. It is sometimes instructive to question our assumptions. Just because the Apollo program did not postulate developing the Moon's resources, (it was never intended to do so) it does not mean that this cannot happen. With even a megawatt or two dedicated to metallurgy, literally tons of iron and oxygen can be generated *per day*. Vacuum induction furnaces similar to what are used in the high tech metals industry on the Earth could be used to melt the regolith (lunar soil) and generate these metals which, because they were made in a high temperature process, can then be poured into molds, rolled into sheets, and cast into objects of utility or desire.

Three feet of iron has equivalent shielding power to the Earth's atmosphere, and large air tight structures could be built on the moon with lunar derived metals. Large lunar bases can be built this way by humans with robotic assistance using mostly local materials. Robots will be a vital part of this future. They were primitive (most still are), and up until the early years of this century, were not that useful outside of very strict environments such as the manufacturing floor. Robotic assistants could be used for a lot of the outside work in the radiation environment guided by human tele-operators or used autonomously. A tremendous amount of work is going on in the development and deployment of these robotic assistants in underground mines and strip pits on the Earth to drive trucks, move large quantities of dirt and do the actual mining. A lunar base could help to improve the state of the art in robotics in the U.S. where we are lagging our Japanese and European friends.

Beginning in chapter 2 I will develop these arguments in detail, beginning with a history of how energy and the inventiveness of humans, has improved our lives and enabled the world's first global civilization. From there we will look at the state of our energy supplies today and the difficulties that lie ahead if we do not find ways to diversify our sources of energy. In chapter 4 I will address the thesis of those who postulate that we face intractable "limits to growth," their solutions and why they are unacceptable.

The story will continue with a look at solutions for energy supplies—from solar to the inexhaustible potential of nuclear fusion. It will then go on to examine the history of the internal combustion engine and the contrasting power of fuel cells along with the key factor that will enable or retard the hydrogen economy, the cost and the limits of terrestrially derived platinum. In chapter 8 the fact that we have always used platinum from extraterrestrial derived resources will be developed as well as the existence of these resources today in the asteroids and on the Moon. Chapter 9 will detail the remote sensing of the Moon in the past and its limitations for discovering PGM concentrations. Chapter 10 through 12 will illustrate past lunar architectures and their fate. In chapter 13, the vision for the return to the Moon will be presented followed in chapter 14 by a discussion of the Aldridge commission and a concept for lunar exploration. Chapter 15 will discuss recent NASA efforts and their applicability to a commercial lunar architecture. Chapter 16 will close with a look at a commercial lunar base and some thoughts for the future.

Thank you for taking this journey with me. The thoughts that I present to you, the reader of this book, have been developed over the past decade and a half of watching the hopes and dreams of space advocates and entrepreneurs grow and ultimately be frustrated by ill formed policies and plans. It is my desire that this book help frame the argument in terms that will appeal to a wider audience of people and to show that the return to the Moon is front and center of how we will solve the key problems of limited resources on our Earth. We are not doomed to live in a resource scarce world, denuded by a rapacious few centuries of prosperity; that is unless those who make the decisions ignore solutions that are as obvious to our generation as the Moon in the sky.

*　　　　*　　　　*　　　　*　　　　*

[i] N. Ruzic, The Case for Going to the Moon, G. P. Putnam Sons, New York, NY, 1965, P. 8
[ii] Science, # 91,620 as quoted by Ruzic, P. 8

Chapter 2:

The History of Energy and its Impact on our Lives

Abundant energy is one of the five issues listed in the previous chapter that defines a great portion of our quality of life for all of us in our civilization. If there are resources on the Moon, or products can be derived from lunar resources that can free us from our dependence on oil, then that would in and of itself justify a return to the Moon. In order to show what these resources are and how they can make a contribution that justifies the cost of going to the Moon, a bit of the history of energy must be understood. A chain of logic must be constructed to make the connection between the Moon and our Energy future. All of the links of this chain are still not widely known, nor is it really taught in school the story of how we have gotten to our present energy situation and its importance in our lives. By understanding where we have been and where we are going concerning energy, we can better see what the key factors of future energy production and use are, and how lunar resources can fit into our energy future.

The History of People and Energy

Since the dawn of civilization until the 1700s, energy production and utilization has been pretty much the same. Wood, water, wind, and the sun were the prime sources of energy for cooking, cleaning, transportation, and agriculture. Muscle took these energy sources and turned them into food, clothes, and articles for trade. The vast majority of the time the muscle used was involuntary, the fate of the animal or the slave to serve the needs of the many and the wealth of the few. Slavery was profitable and common in this energy starved world. For a few brief moments in Greece and Republican Rome, freedom came, but only to a small portion of the total society. In a muscle bound world, slavery had a tremendous economic value to the masters, and no one in that era seriously suggested its end.

In the thirteen hundred year span from the fall of Rome to the beginnings of the industrial revolution life was mostly short and hard for slave and freeborn alike. Average life expectancy in Britain and Europe was 36 years during the 1700s and was as low as 32 in earlier centuries. Not too many baby boomers would be around now if this were still the case. People's attitudes were steeped in a static world and were at the mercy of their lords, kings, and a general climate of "as today was, tomorrow will be." In most countries people were even legally forbidden to move from their hometowns. Violators faced punishments ranging from beatings to death.

There were also times like the era of the Black Death in the 1300's that, if repeated today in like scale, would kill billions. In England, the rest of Britain and Europe, the majority of people living during that era were agricultural workers living on the land.

Some were serfs, with lives only a little better than slaves who tilled the land for a lord. Some were freeborn men who had the use of the common land of the area and were not directly beholden to their lords. Others were yeoman farmers that had what we would think of as a family farm. Then there were the few great lords and the lands owned by the Church and the feudal lords, where the only wealth that existed served the few that had it. During all this time wood was fuel and water was power for small-scale industrial machines as well as for iron and other metalworking. The situation was not all that different in China at the time.

This began to change as Britain and Europe began to grow in population and trade during the mild climate of the Medieval Warm Period and began to accelerate during the Little Ice Age as climate change forced technology forward. The island of Great Britain is not all that large. The land area is slightly smaller than the state of Oregon! However, in Great Britain several things were happening that put that country on the road to being the prototype of the modern world.

Britain was blessed by an abundance of the elements that materially enable an early industrial civilization. In the Midlands and Wales, iron and coal were plentiful. It was therefore inevitable that coal would be used more because wood started to become more expensive and scarce as the forests of Great Britain were progressively cut down. Coal burns hotter than charcoal from wood, and when coal is turned into coke by heating in an oxygen free environment at 1000 degrees C for 48 hours, it makes an even hotter fire. The Quaker iron master, Abraham Darby first used coke for metalworking in England in 1709. It was used to increase the temperature of the iron until it actually melted and could be cast into forms; something rarely accomplished in ancient times. This helped to improve the quality of iron products while the increased productivity from the use of this more efficient energy source lowered prices for finished goods. Lower prices increased demand for these products, which in turn increased the need for coal and iron, leading to increased jobs in mining, metalworking, and manufacturing.

Britain also had an advantage in their economic system. Unlike Europe there were no internal taxes or tolls between different regions of the country. Coal from Wales and iron from Scotland was treated equally in the markets of London, Birmingham, and Manchester. Europe was still mired in the old ways with a patchwork quilt of rules, regulations, and guilds that stifled trade and commerce. A spirit of entrepreneurialism also existed in Britain that was actually encouraged by the state. This was learned in previous centuries from the city states of Italy during the crusades when the King learned that the entire wealth of his country was less than the gold in just one of the Florentine banks.[i] Britain, as an island nation, understood trade and after the fall of Constantinople and the discovery of the new world her people knew that it was through trade and manufacture that the gold stolen by Spain could most efficiently be transferred into British pockets.

Britain is also blessed with an extensive river system. Most of the early industrial sites were located on these rivers because water wheels provided rotary motion, which allowed them to be hooked to gears that would then run machinery. This advantage extended to transportation as well, because roads in England in the 1700s would not have been any better than logging roads are today! As the industrial revolution gained speed, government chartered partnerships and stock corporations built canals that linked these various rivers in order to increase the flow of commerce. When the Trent and Mersey canal was completed, transportation costs for the famous Wedgwood pottery decreased to 1/8th of its previous cost and breakage was greatly reduced by smooth water versus rough land transport of fragile pottery. As other canals were completed, the cost of food, coal, and other commodities dropped by similar amounts. This improvement in transportation efficiency was as instrumental as the invention of industrial machinery and the use of coal in enabling the industrial revolution to move forward, improving the lives of the poor as well as the rich.[ii] Transportation, energy, and innovation do go together and this will be crucial as well to enable the development of the Moon and its resources.

The Age of Steam

All of this change accelerated when the Scotsman James Watt, with the help of industrialist Matthew Boulton, invented the world's first practical and efficient steam engine. The standard up to that time, the Newcomen steam engine, invented in the early 1700's was very inefficient and was only used as a pump to extract water from the coal mines. Watt received a patent for his steam engine in January of 1769, *"for a new method of lessening the consumption of steam and fuel in fire engines."* The first application of Watt's steam engine was for the mines. It could pump four times the water for the same amount of coal as the Newcomen engine. This was just the first of Watt's innovations. With the Watt engine's gain in efficiency, sales increased dramatically and the factory of Matthew Boulton was soon busily expanding to cope with the demand. Most of the machines in the early years went to the mines for pumping water, and in just a few years, displaced the earlier, less efficient engines. Some went to support rolling machines and blast furnaces. Some went to run power looms for the manufacture of clothing. All of these machines turned the wheels of progress and greatly increased the use of coal for energy.

A second, less celebrated, but no less important, patent was given to Watt in 1781 for a method of translating the back and forth motion of a piston in a steam engine into a rotary motion (what we now know as a crankshaft). With this innovation power looms could be run with a single steam engine and a host of other industrial processes automated to improve productivity and production rates, which lowered costs and drove an increase in demand in an ever expanding virtuous cycle.

This second patent on reciprocal-to-rotary motion, coupled with the steam engine, enabled the industries to move close together into cities that, until then, had to site their plants in rural districts near fast flowing water, thereby improving access to workers and transportation nodes. This is the beginning of the factory districts that became the centers of industry in the following years. Coal and steam powered the increasing pace of the industrial revolution. It enabled British industry (Scotland, Watt's homeland was fully integrating itself into the English economy) to out produce any competition from abroad and propelled the exponential growth in British power and wealth in the 18th and 19th centuries. It also improved the lives of citizens and increased life expectancy from approximately 36 years in 1700 to about 40 years during the mid 1850s, and up to 48 years by the year 1900. This was principally due to the improvements in transportation that greatly lessened the impact of local crop failures, and improved access to meat from distant lands such as America.

The use of coal and the invention of steam were not the only inventions that enabled the industrial revolution to happen, but without them, it would have been impossible to have realized the great improvements in productivity that drove the entire cycle. History books are blasé about how this occurred and too often speak of the "Industrial Revolution" as something that just happened. However, without plentiful energy, a system of law that protected the rights of inventors, relatively low taxes, a fairly free society, and a spirit of optimism and entrepreneurial vigor, the industrial revolution would not have happened. Laws to support the industrialization of the Moon will also be required by terrestrial governments to support that effort as well as the risk taking of the capitalistic and entrepreneurial classes. Property rights, a cornerstone of western capitalism, must also be addressed to enable legal protection of the developers of lunar resources.

The Steamship

The inventions and transportation progress of the industrial age continued forward. Robert Fulton, along with others began the development of a steam-powered boat that could move against the flow of a river; something never accomplished before. Fulton's interest in steam power for boats began early. This is evidenced by a letter that he wrote in 1794 to the company owned by James Watt and Matthew Boulton, inquiring about the use of a 3 to 4 horsepower steam engine for that application. Interestingly, the letter was filed and forgotten because Boulton and Watt declared that they were not in the boat business![iii] It was not until October of 1802 that Fulton was able to move ahead at full steam when he signed a partnership agreement with Robert Livingston, a New York plantation owner who had been a member of the Continental Congress and one of the framers of the Declaration of Independence. During the time of the partnership with Fulton, Livingston was also one of the negotiators of the Louisiana Purchase. The partnership was for the purpose of obtaining a patent on the design of a steamship that could carry 60 passengers at 8 miles an hour in still water.[iv]

Livingston was obviously a very politically powerful personage, and through his contacts in the New York legislature, in 1798 he was able to obtain a 20-year monopoly for transportation on the Hudson River using steam power.[v] With Fulton's genius and Livingston's connections, Fulton was able to build his boat, the *North*

River, and in August of 1807, launched it on its maiden voyage from Manhattan to Albany, New York. It took 36 hours for the trip and cost $7 dollars. In comparison, the stagecoach, which was the normal way to get to Albany, cost $10 and took 60 hours or two and a half days.[vi]   The Luddites, and some of the competition who understood that the steamship had just made their boats obsolete, attacked this "new fangled" steamship causing Fulton to hire guards for protection.[vii]   Figure 2.1 shows Fulton's *North River* steamship.

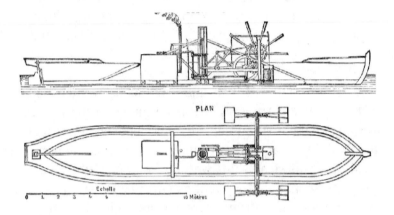

Figure 2.1: Fulton's First Steamship the *"North River"*

Within a few decades water transportation had been completely transformed as this Currier and Ives print shows in figure 2.2:

Figure 2.2: Currier & Ives Print Showing Steam Power
on the Mississippi—Pre Civil War

The boat was an immediate commercial success and the partners cleared $16,000 for thirty weeks of operation. This more than paid for their original investment and would continue to bring in profits for years to come. The reign of steam quickly took over

American waterways, especially the Mississippi as shown in figure 2.2, as the paddlewheel designs pioneered by Fulton stimulated commerce more than anything else in the early years of the nation. Paddle wheeled boats became icons for the American public as well, in song, stories, as well as the frequent disasters from their explosions.

Steam demonstrated the portability of power, and the ready availability of wood and then coal in the United States enabled the nation to grow quickly in the 19th century when roads were scarce and in worse shape than those of Britain. Then British innovation struck again with the advent of the railroad and the first practical locomotive.

It seems that the efforts of the British and Americans went hand in hand in the development of the rail age. At nearly the same time that the American, Col. John Stevens, was working on the first railroad, George Stephenson was preparing to do the same in Britain. It is interesting to note that Col. John Stevens began his earnest interest in railroads after his steamships were locked out of the Hudson River by the Fulton and Livingston monopoly. Stevens is considered one of America's greatest early mechanical geniuses but, unfortunately, his and Fulton's interests did not mesh. In 1812, Stevens petitioned the commission appointed by the government of New York to plan the route of the Erie Canal, to lobby for a railroad rather than the canal. Unfortunately, the two most powerful commission members were Fulton and Livingston! The Erie Canal was built, but within three years, Stevens had been granted a charter by the state of New York to build a railroad. However, nothing more than a small, circular railway was built until locomotives from Britain were first imported to the U.S. in 1829.

In England, it was a race and a prize that launched Britain's railways. Until the invention of the steam locomotive, horses pulled carriages down a wooden or iron railed track. To replace this slow, cumbersome method of transportation, a 500 Pound Sterling prize was to be given by the London and Manchester Railway for the fastest locomotive. Ironically, George Stevenson's entry was named, "The Rocket!" From the website www.inventors.about.com:[viii]

Pass now to Rainhill, England, and witness the birth of the modern locomotive, after all these years of labor. In the same year of 1829, on the morning of the 6th of October, a great crowd had assembled to see an extraordinary race—a race, in fact, without any parallel or precedent whatsoever. There were four entries but one dropped out, leaving three: The Novelty, John Braithwaite and John Ericsson; The Sanspareil, Timothy Hackworth; The Rocket, George and Robert Stephenson. These were not horses; they were locomotives. The directors of the London and Manchester Railway had offered a prize of five hundred pounds for the best locomotive, and here they were to try the issue.

The Rocket won the prize, met all of the requirements of the railroad and was immediately adopted as the British standard. The use of prizes to foster technological innovation is a recurring theme from the 18th through the early 20th century. The Stephenson locomotive was exported to the U.S. and British rail technology became the standard for American railroads as well.

With the advent of a practical locomotive and the increase in iron production, railroads rapidly expanded across the United States, finally bridging the nation almost 40 years later in 1868 with the Herculean efforts of the Union Pacific in the east moving west, and the legendary Stanford, Hopkins, and Huntington's equally hard charging Central Pacific in the west moving east. They met in the desolate stretch of desert north of Salt Lake City Utah called Promontory. The construction of railroads, locomotives, and steamships hastened technology forward along a broad front. Steam shovels were invented to move the massive quantity of dirt and rock excavated for cuts and passes on the railroad. Nitroglycerine was invented to increase the productivity of blasting rock to feed the steam shovels. Iron was used in accelerated quantities. Colis Huntington, of the Central Pacific Railway, ordered, and had shipped over from England, 16,000 tons of iron parts, as well as 24 locomotives in the spring of 1868.[ix] This one order is greater than the entire productive capacity for iron in Britain two centuries before.

When the America's, "National Railroad," as it was called, and the rail system in Great Britain, were completed, economic growth accelerated its pace. Increases in wages and wealth were recorded for most economic sectors. Food distribution between the east and west coast became possible and profitable, leading to better nutrition and increased lifespan. With steamships becoming a standard way of shipping between the U.S. and Europe, trade increased substantially. In general, the time between the end of the American War Between the States in 1865, and the First World War, was a boom time as the industrial revolution gained hold of the economies of the rest of Western Europe.

The Coming of Oil

Coal was the king of the 19th century. By the beginning of the 20th century, coal ran the ships, powered the railroads, heated homes and fed the blast furnaces of industry. Trains carried millions of people in the U.S. and Europe on their daily travels between cities and across the nation. Travel between New York and San Francisco took two months by sea and then across the Isthmus of Panama (decades before the canal was built), or six months by wagon train in 1860. By the completion of the national railroad and the initiation of regular service this was reduced to four days. Other railroads north and south of the first one were built and the beginnings of the American and British modern transportation network became recognizable. However, a discovery made a few decades earlier in 1859 in Pennsylvania would shortly dethrone the king and enthrone a new one.

Until the discovery of oil from underground wells was made, there was not really much demand for oil. The oil from whales was the most widely used liquid hydrocarbon with a wide market and it was used mostly for lamps and lighting. This became a great source of wealth to New England and the foundation of the whaling industry. However, as the 19th century wore on, whales were becoming scarce. At the site of an oil spring in Venango County Pennsylvania, a well was dug by an enterprising group of men to see if the already profitable oil trade from the springs could be increased. The first well hit oil at a depth of 71 feet and produced about seventy-five barrels of oil each day.

With a production cost of about a dollar a barrel and a sale price of six dollars a barrel, there was great profit to be had and, thus, the oil industry was born. Thousands of wells were drilled across Pennsylvania and a wealth that eclipsed that of California's gold fields in terms of economic value was created.[x] In Clarion county Pennsylvania alone, production jumped from 8 barrels a day in 1866 to over 6500 per day in 1877. Costs declined from $6 on the market to as low as 97 cents a barrel before climbing back to about $2.10 in 1877. In comparison to today's cost, this would be somewhere around a hundred dollars a barrel![xi]

The reason for the dramatically increased interest in oil was a discovery by a Mr. James Young of Scotland, that he could distill light oil suitable for replacing whale oil for lighting by "destructive distillation," or what we call refining (or catalytic cracking) of bituminous shale (oil bearing shale). This process was soon extended to crude oil and a refinery was built in Pittsburg for the distillation of the oil from the oil springs in 1850. This was a very successful business and stoked the desire by other entrepreneurs to drill for the oil that was discovered in 1859. Even with the relatively high price of oil then, it was much cheaper than whale oil. This signaled a turn away from whaling for oil, which brought the eventual extinction of the New England whaling industry, but enriched Pennsylvania and saved the whales as a happy by-product.

As more and more wells were drilled in other parts of the United States such as Oklahoma and Texas, supply increased and demand widened in scope to begin taking the place of coal. By the year 1900, oil production had increased to over 149 million barrels per year and this was just the beginning. New inventions would begin oil's meteoric rise in demand. The principal new invention and demand driver was the automobile. Most of the automobile manufacturers that we have today got their start in the last decade of the 19th century and were aided by the availability of gasoline and the invention of the gasoline internal combustion engine by Karl Benz in 1886.[xii] This engine, and the one invented at about the same time by Nicolas Otto, ushered in the automobile age and fueled the several fold increase in demand for oil. Oil production rose from 149 million barrels in 1900 to over 63 *billion* barrels a year in 2000.

Transportation has been only one of our many oil-consuming activities. At about the time that the internal combustion engine was invented and began limited production, the Serbian genius, Nikola Tesla, single handedly invented the other half of our modern civilization: alternating current (AC) electrical power. Tesla's patents in 1888 covered both the generation of electrical power with generators and electric motors to replace steam powered ones. These motors also soon replaced the earlier inefficient DC motors for power generation favored by Thomas Edison. Tesla, along with his financial backer George Westinghouse, brought electrical power to the masses with the construction of the Niagara Falls generating station and inexpensive electrical power became available to the region around the falls.

Before the invention of the Hall process, which required large amounts of electrical power, Queen Victoria in 1868 was given a set of Aluminum tableware for placement in the crown jewels. The cap of the Washington Monument was made from Aluminum as a testament to its value and a tribute to the president. Using the power generated by the Niagara plant, industrial AC power was first used by the Aluminum Corporation of America, in Buffalo New York, to make… what else… Aluminum, using the Hall process. Within forty-five years hundreds of thousands of tons of Aluminum were produced per year and used to build the airplanes that won World War II for the allies. Starting with the Niagara generating plant, and advancing across the nation with TVA and other government and private operators, electrical power soon began displacing other forms of energy in the nation for lights, heating, and industry.

Electrical generation soon moved beyond using just hydroelectric power to using coal and oil fired steam engines and later steam turbines to turn electrical generators. Today we have also incorporated nuclear fission reactors as a heat source for the steam to turn the turbines today and possibly fusion in the future. Over time, electrical power generation became one of the top consumers of both coal and oil, this being one of coal's last remaining competitive uses. Due to the tightening grip of environmental regulations and the perceived public fear of nuclear fission power, power generation units have moved to lighter and lighter forms of hydrocarbon such as liquid natural gas (methane and other light hydrocarbons mixed). Even though the last nuclear plant to be licensed in the U.S. was in 1978, fully 20% of the electrical power in the country is generated via nuclear power. Since the end of the growth of the nuclear power industry, most of the AC power plants built in the past 20 years have been oil or liquid natural gas fired. All of these energy sources power our civilization today. They light and heat our houses, run our washing machines, dryers, televisions, computers, and all the other electrical devices that make our lives so much easier than that of our ancestors.

In so many ways, we who live in the west have benefited from plentiful, inexpensive oil. It is interesting to note that the life expectancy of the general population in England rose from 36 to 48 years, since the beginning of the industrial revolution until 1900 when oil began to take over from coal and wood for primary power. Since the advent of an energy economy based on oil, our life expectancy in Britain and America exceeds 78 years for someone born today. That is more than double the life expectancy in a little over 300 years. This is mostly due to oil, the inventions enabled by its existence and the medical and scientific advances that benefited from it. Energy use permeates our lives today in ways that most of us are oblivious to. An attempt to make that connection was provided recently by a series of reality shows on television that set the participants in various time periods and let them live the lives of pioneers in the west, the Victorian middle class, and suffer war shortages in a WW II home. These people all came away from these experiences with a much greater appreciation of how much easier our lives are today due to the invention of "labor saving" devices.

There are now tens of millions of people under the age of 11 who cannot remember a time before the Internet. The Internet is also fueled by oil. The problem is that we will run out of oil at some time in the future and the majority of the oil that we have left is mostly in politically unstable areas of the world. In the next chapter we will look at oil today, its sources and reserves, and the arguments of those who predict our energy future; who paint either rosy or pessimistic appraisals of where we will be in a hundred years if there is no alternative to oil and what their ideas are to deal with that eventuality. This is where the connection from energy to the Moon and its resources will begin to be explored.

*   *   *   *   *

[i] I. Montanelli, R. Gervaso, *Italy in the Golden Centuries*, Henry Regency Company, Chicago, 1967, P. 121

[ii] J. Uglow, *The Lunar Men,* Farrar, Straus, and Giroux, New York, NY 2002, P. 116-188

[iii] K. Sale, *The Fire of His Genius*, The Free Press, New York, NY, 2001 P. 57

[iv] Ibid, p. 88

[v] Ibid, p. 33

[vi] Ibid, p. 129

[vii] Ibid, p. 115

[viii] http://inventors.about.com/cs/inventorsalphabet/a/oliver_evans_3.htm

[ix] D. Bain, *Empire Express*, Penguin Books, New York, NY, 1999, P. 450

[x] http://www.pa-roots.com/~clarion/books/caldwell/oil2.html

[xi] Ibid

[xii] http://inventors.about.com/library/weekly/aacarsgasa.htm

Chapter 3:

Energy Today and Our Options for the Future

We live in what is arguably the most incredible time in all the history of humankind. Today a significant portion of the planet's population lives at a level that would be the envy of a Roman emperor and the majority of people on the earth are far better off than our ancestors. From our modern entertainment, labor saving devices, our modern freedom of travel (seventy five million passengers flew on just Southwest Airlines last year), to the increased access to food from around the world that mitigates the effects of crop failures and seasonal fluctuations, more people than ever in history have a high standard of living. Table 3.1 gives some of the mind-boggling statistics regarding human travel just in the U.S. as of the year 2001.[i]

Transportation Method	Millions of Miles Traveled	Last Year of Survey
Air (Passenger miles)	531,239	2000 (down in 2001)
Automobiles	4,432,327	2001
Mass Transit	490,070	2001
Railroads	5,559	2001

Table 3.1: U.S. Total Passenger Travel Miles for Year 2000

These are numbers that would have been considered magic three hundred years ago. Each year, Americans travel farther than the entire human race did in the thousand years before 1700. As Americans, we take these numbers in stride. This is all a result of access to plentiful energy, particularly oil. Plentiful energy has helped to increase life expectancy and it has helped to shatter the bounds of slavery (mechanization dramatically reduced the value of chattel labor), and has the potential to lead mankind to the Moon, to Mars, and beyond to the stars.

Today we think nothing of buying cars from Japan, electronics from Asia, wine from France and a plethora of industrial and convenience items from nations as far flung as Australia to Russia. The next time you go shopping, look at the country of origin of many of the items that you buy on a daily basis. For the first time in history, we have a tightly coupled global economy. The U.S. also ships enormous quantities of finished goods as well as raw materials around the world. While we often hear about the trade deficit, we rarely hear the full story about the hundreds of billions of dollars worth of our products that get shipped overseas. Trade makes all of our lives better and has the potential to continue to do so except that…..

This will all go away unless we can maintain and even increase the amount of energy available to our civilization. Without plentiful, portable energy, our standard of living will fall radically and we will not be able to maintain the intricately interwoven threads of a global economy that has brought us the benefits that so many of us take for granted today. Therefore, we need to increase the amount of energy available so that everyone who wants to live at this level can do so, without destroying the environment, or leading to global collapse from using more resources than it is possible to obtain from the earth. The question is how can we do this with a supply of oil that is far from infinite. The pessimists say that this is impossible, that technology cannot overcome the limits to growth that we face today.

Recently, a World Wildlife Federation press release proclaimed that it would take the resources of two more Earth's to enable the rest of the world to rise to a western standard of living. They are probably right and one of those planets is visible in the sky above our heads every day. While the Moon is not a planet as the Earth is, it does have resources that can be used to overcome the limits to growth that a one-planet civilization such as ours has today. The Moon is also a springboard to the other planets and Moons of our solar system that have vastly greater resources than our Moon. Forging the links of our chain of logic continues here with a review of the state of our supply of energy today.

The End of Oil

As of 2003, the United States uses about 15.5 *million* barrels of oil *per day*. Out of this total, almost ten million barrels of oil and over two million barrels of refined products are imported each day. At today's price of $38 per barrel, that adds up to about $367 million dollars per day that we send out of the country for energy (not including the two million barrels of refined products). Each year, this is over $134 billion dollars. Today we only domestically produce about 36% of the oil that we consume.[ii] This situation is not going to get better with time as the amount of oil that exists on the earth is clearly finite and we are already near the all time peak of production.

We are running out of oil. It is really that simple. We may not run out tomorrow and we may not run out for fifty years but within the lifetime of most people alive today, we will substantially run out of oil to the point that maintaining civilization at its current level will be impossible. Unless we find new sources of energy we can count on our standard of living falling, our days of driving SUV's ending, and a lot of bad days ahead for the human race. In this chapter, statistics concerning oil and our reserves, our uses, and current projections by the U.S. government, as well as by the global pessimists and optimists, will be presented. However, no matter how optimistic the scenario, the year 2100 will find our civilization either with a new source of energy and energy utilization, or we will find our civilization dramatically poorer than it is today. The latter is not a foregone conclusion. The energy is out there, orders of magnitude more than we have available on the Earth today. This is where space, and particularly the Moon and its resources, has the potential to bring us through the coming transition in a way that leads to a prosperous future.

The State of Oil Supplies Today

There are broadly speaking, generally three consensus group opinions on the amount of global oil supplies remaining. The first group I call the global optimists. This group is led by the Saudi government, their oil company Aramco, and most of the rest of the oil industry. The second group I call the global pessimists. This is nominally led by environmental groups, as well as some of the best experts in the oil industry itself, and both these often-conflicting political groups see a dangerous complacency in the "no-worries" predictions of the optimists. In the middle of these two extremes are the government agencies that have the responsibility for forecasting the future of energy supplies. The estimates that the U.S. government have developed are somewhat more conservative than the global optimists but far less pessimistic than the global pessimists. Where is the real story? Let's see what the data says and then go from there.

The Global Optimists

The leading example of the global optimists, the Saudi Aramco oil company consistently estimates a high level of remaining oil. In a presentation in February of 2004, Mr. Mahmoud M. Abdul Baqi, Vice President of Oil Field Development from Saudi Aramco, gave a presentation on *"Fifty Year Crude Oil Supply Scenarios: Saudi Aramco's Perspective."[iii]* In this very informative presentation, Mr. Abdul Baqi presented a vast amount of data concerning the state of Saudi reserves (the largest reserves in the world, representing 25% of the global total), and the capacity of Saudi oilfields to maintain production at high rates, as well as eventual depletion scenarios. It is striking that in none of these scenarios does their reserves last at today's production capacity beyond the year 2054. From the report:

> *World energy demand is expected to increase at an annual rate of 1% to 2% over the next 15 years, reaching an annual demand of 107 million barrels per day by 2020, partly as an anticipated consequence of growth in China, India, and other South East Asian economies.*

> *Worldwide oil reserves at year-end 2002 stand at 1050 billion barrels, of which 65% (or 686 billion barrels) is in the Middle East, with Saudi Arabia being the principal player. The Middle East contributes about a third of total*

*world production, has reserves-to-production life of 92 years and is expected
to play a pre-eminent role in the global energy theater.*

Table 3.2 below gives the oil reserves for the world according to the Saudi report:

Region	Oil Reserves (Billions of Barrels)
North America	50
Central and South America	99
Africa	77
Europe	19
Former Soviet Union (includes republics)	77
Far East	39
Middle East	686
Total	*1047*

Table 3.2: Estimated Global Oil Reserves From Aramco Report

This is where things get interesting. According to their own data, world oil
consumption is 75 million barrels a day or 27.393 billion barrels a year. At this
consumption rate, global supplies will last about another 38 years at the current rate
of production. That is the bad news. The good news is, according to the Saudi's, there
is considerably more oil than what the current reserves imply.

Over the last twenty years, *"Exploration, delineation, and development efforts have
increased Saudi Aramco's oil initially in place by 17% during the past twenty years
(1984-2003)"*. This implies that over the next twenty years there will be an additional
200 billion barrels of oil discovered and added to their reserves. What does this mean?
It means that since they discovered 17% more over the last 20 years that they expect
to discover 17% more over the next twenty years giving them approximately 900
billion barrels of total reserves in place. Oil Initially In Place (OIIP) has a complex
meaning. Following is the definition of that term:[iv]

Oil Initially In Place
Reserves: Oil that can be recovered commercially with current technologies
Proven: Greater than 90% certainty
Probable: Greater than 50% certainty
Possible: Greater than 10% certainty
Contingent Resources: Less than 10% certainty

Here is what is interesting and gives the global pessimists and the U.S. government
middle of the road estimates their argument. According to the same document that the
above information comes from, the Saudi reserves are as follows:

Proven	260 (Commercially recoverable at 90% certainty)
Produced	99 (This has already been pumped!)
Probable	32 (greater than 50% certainty)
Possible	71 (greater than 10% certainty)
Contingent Resources	238 (less than 10% certainty)
Total	~700

Table 3.3: Saudi Oil In Place Estimated Reserves

The above estimates are not reassuring. The Saudi Aramco data indicates that out of the 686 billion barrels, 238 billion of that is speculative. On top of that, the additional 200 billion that they estimate finding over the next 20 years is speculative as well. In the U.S., the Securities and Exchange Commission (SEC) would not allow this type of speculative accounting to be listed as an asset. This is exactly what has gotten Shell Oil in trouble recently when reserves in the neighboring Sultanate of Oman were found to be considerably overstated. In an article from the New York Times[v] by Jeff Gerth dated April 8th 2004, it was stated:

The declines in the Yibal field are spelled out by officials of the joint venture in two papers that were published last year by the Society of Petroleum Engineers. The papers have different numbers: both say production peaked in 1997, but one said it declined to its current rate of 88,057 barrels a day by 2000 from a peak of 251,592, while the other said it fell to 95,000 barrels from 225,000. A spokeswoman for the society said she could not explain the difference.

The most disturbing aspect of this misstatement is that Shell is one of the few joint ventures between a western company and an OPEC country in the Middle East. What this means is that this is the only insight that we have into what is going on in Middle Eastern oilfields that does not come from a government owned company that must (theoretically) accurately report changes in their proven reserves and they are dramatically downsizing their reserve estimates. So far over the last several months of 2003 and 2004, Shell has reduced the amount of global reserves on their books by 40%.

In an April 5, 2004 article in Business Week magazine by Stanley Reed, less than two months after the presentation by Abdul Baqi referenced here earlier, doubt is being raised about the quantity of Saudi oil reserves as well as their ability to continue pumping oil at high rates. The Saudi's maintain that they can keep their pumping capacity high and said:[vi]

The Saudis conclude that the kingdom could easily ramp up to 10 million bbl.[barrels] a day from its current 8.5 million and comfortably sustain that level through 2042. If demand is really strong, they insist, the kingdom could build up to 12 million bbl. a day by 2016 and hold that level out of existing reserves until 2033.

The disturbing part about this pronouncement by Mr. Baqi in the article is that the amount of oil that he is saying that they can pump is less than 60% of what they claimed in the *"Fifty Year Crude Oil Supply Scenarios"* presentation just two months before. Also, in the presentation it is claimed that the rate of oil production can be sustained at 12 million barrels a day until 2054. Clearly, there is a disconnect between one and the other statements by Aramco.

The most ominous thing brought out in the Business Week article is that there is no "Plan B." In the words of Matthew R. Simmons, chairman of Simmons & Co. International, a well-known Houston investment bank:

> *"The entire world assumes Saudi Arabia can carry everyone's energy needs on its back cheaply," says Simmons. "If this turns out not to work, **there is no Plan B**."*

The article points out that other oilfields that were discovered at about the same time as Ghawar in Saudi Arabia (1948), such as Forties in Europe's North Sea, or Alaska's Prudhoe Bay, have all suffered dramatic declines in productivity. According to the Saudi report, this decrease in the non-Saudi oil fields is as much as several percentage points per year.

The optimistic scenarios that are being painted about reserves from the Saudi Aramco oil company may be just that, optimistic. This leads us to the estimates from the U.S. government that represent (as of 2000) the best estimates of global reserves.

U.S. Government Estimates

The U.S. government has extensive public documentation of information relating to oil reserves. There is so much information that it is somewhat difficult to wade through the data to find out what is important. Also, there are differences in the numbers due to the different ways that the U.S. government and foreign sources account for different types of reserves. Add to this the fact that there is no objective standard that everyone follows even for foreign data compilations. In the Saudi presentation (Table 3.2), the number they use for North American reserves, which is 50 billion barrels, would be considered proven reserves. This is compared to the 260 billion barrels of oil from Saudi Arabia and 686 billion barrels across the entire Middle East. Obviously North American reserves are only a fraction of those remaining in the Middle East.

Table 3.4 gives the U.S. government statistics for U.S. reserves:[vii]

Technically Recoverable	Crude Oil (Billions of Barrels)	Natural Gas (Dry Trillions Cubic ft)	Natural Gas (Liquid Billions of Barrels)
Known	67.70	390.00	13.40
Possible (5%)	174.82	1430.63	23.67

Table 3.4: U.S. Government estimates of Proven and Probable Oil and Gas Reserves

The proven reserves are somewhat higher than the Saudi estimates because the U.S. data includes undiscovered reserves (reserves that they think are there due to remote sensing and geology, but not proven). It is interesting to note that the estimated reserves in the Artic National Wildlife Refuge (ANWR) that has been so controversial in congress, comprise 15% of our total estimated reserves. All of this information is available online at the U.S. Department of Energy's Energy Information Administration's website at www.eia.doe.gov.

Forecasts for oil reserves around the world outside of the United States vary widely. A file on the DOE website (AR.pdf) shows several hundred percent difference between the 95% probability of oil being found (pretty well known) down to the 5% level (a hope and a prayer).[viii] The document reads like this:

> *The world potential reserve growth (excluding the U.S.) for oil is estimated to range from 192 BBO at a 95 percent probability to as much as 1,031 BBO at a 5 percent probability, with a mean of 612 BBO [billion barrels of oil] (fig. AR-27). These estimates, as well as those for gas and NGL [natural gas, liquid], reflect a 30-year forecast span (1995 to 2025).*

The difference between oil reserves that are pretty well known and those that are mostly speculative is huge. This difference does not inspire confidence, especially since most of the difference is in the Middle East, a region rife with uncertainty in the quality of our available information. However, the document does have a very good display of the information on a country-by-country basis regarding known and unknown (between "pretty good" and a "hope and a prayer") reserves. A summary of this information is provided in Table 3.5 below:[ix]

Region	Pumped Oil	Known Oil	Undisc Oil	Pumped Dry Gas	Known Dry Gas	Undisc Dry Gas	Pumped LNG	Known LNG	Undisc LNG
OPEC	280654	853132	281176	119410	2229901	1609358	2021	41326	89411
OECD	45441	77371	29250	242542	640821	436965	2795	12568	19339
Other	213092	466337	338210	533470	2648960	2623147	2567	21802	110794
USA	165800	67700	174820		390000	143063		13400	23670
Total	*704987*	*1464540*	*823456*		*5909682*	*4812533*		*89096*	*243214*

Table 3.5: U.S. Government Estimated Global Oil Reserves (Proven and Possible)

There is a wide divergence between estimates that seems to be based upon the level of confidence in data. The data above is in the middle of most of the estimates that my research has uncovered. There is some fairly wide uncertainty related to additional possible reserves of oil and gas on the outer continental shelves.[x] Basically, beyond the "known oil", the undiscovered oil, gas, and Liquid Natural Gas (LNG) columns above are based upon statistical evidence and the validity of that evidence is at best speculative (*Undisc* above is for undiscovered)

What this all means is that, between the optimists represented by the Aramco report and the U.S. government report, if you add all of the oil in Table 3.5 and divide by the average forecasted daily demand (2.29 trillion barrels divided by 36 billion barrels per year [100 million barrels per day] oil completely runs out in 62 years. Long before complete depletion, the peak amount of oil that can be pumped will decline. This is called the Hubbert peak, from the oil field engineer that first did a study on how long it will be before the maximum rate that oil can be extracted begins to decline. One thing that the above numbers make absolutely clear, the amount of oil and gas that remains to be discovered is less than what we know of today based upon the optimistic statistics presented here. This inevitably leads to the end of the oil economy well before the year 2100.

The Global Pessimists

In an article recorded on CNN.com in October of 2003, it was reported that a group of scientists from Sweden, led by Professor Kjell Alekett, have come to the conclusion that oil production levels will peak in about ten years time and that current estimates of the world's oil reserves are only about 20% of what actually exists.[xi] The article makes the statement:

> *Alekett said that his team had examined data on oil and gas reserves from all over the world and we were "facing a very critical situation globally." "The thing we are surprised of is that people in general are not aware of the decline in supplies and the extent to which it will affect production." "The decline of oil and gas will affect the world population more than climate change."*
>
> *According to the Uppsala team, nightmare predictions of melting ice caps and searing temperatures will never come to pass because the reserves of oil and gas just are not big enough to create that much carbon dioxide (CO_2).*

Alekett said that as well as there being inflated estimates of probable finds, some countries in the Middle East had exaggerated the amount of reserves they had.

A lot of these pessimistic estimates can be traced back to a Dr. Colin Campbell, a petroleum consultant who has written books and papers on the subject, and now is working with the M. King Hubbert Center for Petroleum Supply Studies at the Colorado School of Mines. In a document called, "Forecasting Global Oil Supply 2000-2050," Dr. Campbell uses these numbers for oil production in Billions of Barrels illustrated in Table 3.6:

Oil Reserves	World	Middle East	Russia	USGS Data
Produced	875	225	120	704
Reserves	900	500	70	1464
Undiscovered	150	40	15	823

Table 3.6: Hubbert Center Global Oil Reserves Estimates (Plus the USGS Data)

While the numbers for reserves are similar to the USGS data, the undiscovered reserves are only a fraction of what the USGS and the Saudi's estimate. There is a good reason for this and it is illustrated in this excerpt from Dr. Campbell's Hubbert report.[xii]

For example, in an unknown, untested, basin in East Greenland, it concluded that there was a 95% probability (F95) of finding more than zero, namely at least one barrel, and a 5% probability (F5) of finding more than 112 billion barrels. A Mean value of 47 billion was computed from this range. Since the numbers were quoted to three decimal places, the reader could be forgiven for assuming them to be accurate. But a moment's reflection would question the very concept of a subjective 5% probability. In plain language, it was a guess that could as well be the half or the double, yet it entered the calculations distorting the critical Mean value. We are now seven years into the study period and can compare the forecast with what has been found in the real world. The USGS forecast, as a Mean estimate, that 674 Gb (billion barrels) are to be found between 1995 and 2025, which means an average of 25 Gb a year. So far, the average has been only 10 Gb, when above average performance should be expected because the larger fields are usually found first because they are the biggest targets.

This means that the picture drawn from the US Geological Survey report, and the Aramco report, is well biased toward the "hope and a prayer" end of the spectrum of oil discovery. This is further illustrated by a graph of oil discoveries from 1950 through 2000, followed by the Hubbert Report projections:

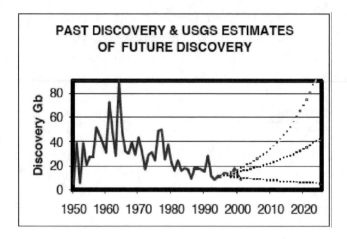

Figure 3.1: Past Discoveries and Future Estimates (From the Hubbert Report)

For those who are used to looking at graphs of data, Figure 3.1 reveals a disturbing trend. The dotted lines represent three different scenarios related to future oil discoveries. The top line would be a very optimistic scenario, interestingly close to one that the Intergovernmental Panel on Climate Change (IPCC) endorses. This was revealed in the CNN.com article where the IPCC estimates oil reserves at up to 18,000 billion barrels, which is well outside of the realm of possibility but forms the basis for a lot of global warming research. The middle line represents the Aramco and USGS scenarios. The bottom dotted line represents the 95% confidence level (Proven reserves in Aramco terminology). The truth is probably between the lower of the two lines, less optimistic than the USGS but more optimistic than what Dr. Campbell estimates.

Using this as a basis, or even the more optimistic scenario of the USGS, there is a big difference between the eventual depletion of oil and the peak of oil production. Campbell speaks of this:

> *The critical issue is not so much when oil will eventually run out, but rather when production will reach a peak and begin to decline, which will represent a major watershed for the world's economy.*

In the presentation by Aramco, their estimate is that the peak of oil production, even at a rate of 12 million barrels per day, would not happen until 2054. However, this is predicated upon the most optimistic assumptions, and even Aramco only commits to 2033 as the end of their production plateau in interviews with the popular press. The trouble with the world economy will come far before the depletion of oil; it will begin with the end of the plateau of oil production (which means that beyond that point, whatever production level exists at that time can no longer be maintained). According to Campbell, the end of this production plateau could come as soon as 2010. While he has set his target as early as 2005 and was wrong, clearly this is a matter of global concern and should be debated and addressed in a forceful manner.

Moving over to natural gas only delays the problem a few years. It would be difficult to transition the entire global automobile fleet to this fuel source and with that level of demand, Liquid Natural Gas (LNG) would be rapidly depleted.

So, what do we do? This will be dealt with in chapter five, but as a preface, there are two broad paths that present themselves for the consideration of the reader and the global politic. One is the path of the movement I call, "The Limits to Growth Movement," or LGM, that is based on the book, *"Limits to Growth,"* published in the early 70s, that has been the standard reference guide of the environmental movement. This group is represented by no less a personage than former Vice President Albert Gore, author of, *"Earth in the Balance,"* the 1990's version of, "Limits to Growth." Also, in 1992 the successor to "Limits to Growth" was published. Called *"Beyond the Limits,"* and was written by the same authors, who in the successor book declared that we are "beyond" the limits that they declared twenty years previously. Neo-Luddite patron, Kirkpatrick Sale, also falls into this group with his book, *"Rebels Against the Future."* Broadly stated, the groups that fall into the category of anti-globalization also are a part of the LGM movement. The middle ground is, in practice, the status quo that right now is pretty much ignoring the whole issue. This includes congress as well as American and global industry. The other side of the equation (you knew this was coming) is the solution represented by space advocates, and in particular, the effort to return to the Moon and develop the rest of the solar system. This is the only group that actually has an idea to use the resources of the Moon to address the problems that we have on the Earth.

There are other factors that have to be taken into account regarding oil and energy that cross over into all of the other global concerns enumerated in chapter one. Probably the most well-known and debated concern today about energy from oil is the effect that the combustion of hydrocarbons has on the atmosphere. This is known popularly as "global warming," or the CO_2, or the "Greenhouse" problem. This also will be further developed in the next chapter, but an interesting statement as a result of the global pessimists argument is that, we will not have to do anything about the CO_2 problem because, with oil running out, there is no supply of oil to continue to drive the CO_2 concentrations to the level required to sustain the rise in temperatures estimated by the Intergovernmental Panel on Climate Change!

The Hydrocarbon Wild Card

There is a wild card in this hydrocarbon story and it is in the form of methane hydrates. These hydrates are speculated to be derived from the activity of living organisms beneath the sea floor on the continental margins of many areas of the world. They are also found in areas of the Artic underneath the cover of the permafrost. Conservative estimates are that the amount of methane hydrates are between 2 to 10 times the amount of oil and gas in the world![xiii] Now for the bad news, these methane hydrates are mostly in a very diffuse form, akin to other diffuse oil resources in shale and tar sands.

According to the paper referenced here by Dillon and Max, while most of the methane hydrates are in diffuse, hard to extract form in deep sea floor sediments, a significant amount of these gasses, *"appear to exist in significant concentrations."* There is still much to be learned about these resources but interestingly, much of them lie off the continental coast of the U.S. and Japan. If these resources can be accessed in large quantity it could change the balance of economic power, extend our reserves of hydrocarbons, and provide the bridge that we need to enable us to transition to the full hydrogen economy. These methane hydrates will be much more valuable as a fuel for fuel cells rather than for internal combustion engines. No matter what our energy future is, oil will not be a part of it for much longer, and internal combustion engines run on that fuel. So let us continue on forging our chain and see what the various ideas are for our future and where the Moon fits into the plan.

*   *   *   *   *

[i] http://www.bts.gov/publications/national_transportation_statistics/2003/html/table_01_34.html

[ii] Energy Information Administration, Petroleum Supply Monthly, December 2003

[iii] http://www.saudi-us-relations.org/energy/saudi-energy-reserves.html

[iv] Petroleum Reserves Definitions, SPE/WPC (1997)

[v] http://www.nytimes.com/2004/04/08/business/08OIL.html

[vi] http://www.businessweek.com/magazine/content/04_14/b3877007.htm

[vii] http://www.eia.doe.gov/emeu/international/petroleu.html#IntlReserves (file appg.pdf p5)

[viii] http://www.eia.doe.gov/emeu/international/petroleu.html#IntlReserves (ar.pdf p 36 U.S. *Geological Survey World Petroleum Assessment—2000 Assessment and Results*) as of April 2004

[ix] Ibid P 180-183 Table AR-9

[x] http://pubs.usgs.gov/dds/dds-060/sum4.html#TOP

[xi] http://edition.cnn.com/2003/WORLD/europe/10/02/global.warming/

[xii] http://www.hubbertpeak.com/campbell/ (Campbell_02-03.pdf)

[xiii] W. Dillon, M. Max, Gas Hydrate in Seafloor Sediments: Impact on Future Resources and Drilling Safety, OTC 13034, Offshore Technology Conference, Houston, TX, May, 2001, P. 2

Chapter 4:

Energy, Resources, and The Search for Solutions

Countdown to Decision

Today, there is a global debate underway that encompasses potential solutions to the energy problem as well as the interrelated problems of general resource depletion, pollution, and population. I want to keep the focus on energy because the other problems are solved by default either in a good way or bad way depending upon the energy solution and how it is implemented. The frightening thing about this debate is that, while it is public, the public is mostly either unaware or unconcerned. The terms of the debate seem to be framed only by those who are activists for their own solutions; an example of the competing interests discussed previously in chapter one. This is unfortunate. As long as the activists do not align and make common cause with "competing interests" who share broader goals, it is likely that what they propose, if they are able to get it implemented, while workable in theory, would not be in the best interests of the people of our civilization.

Some of the solutions being advocated may very well shock the reader. We live in an information saturated age, and one of the miracles of the Internet is that much of this information can be verified and further researched with nothing more than time and a good search engine. While it is difficult to decipher what some of these groups are actually saying, what is important to know is that the main points are similar to each other. In the end, the presentation of a couple of examples here is adequate to convey the general direction that many of them want to go. In this chapter, I will discuss some of these solutions and their ramifications before going on to present the solution that incorporates the resources of the Moon and the possibilities that this path represents.

"The Limits to Growth"

In April of 1968, a group of scientists, educators, economists, humanists, industrialists and civil servants, gathered together in Rome at the invitation of Dr. Aurelio Peccei, an Italian industrial manager and economist, to discuss, *"a subject of staggering scope—the present and future predicament of man."* As a result of this meeting a group was formed, called, "The Club of Rome," to address the state of the world and to develop possible solutions to problems facing civilization.[i] They were governed by:

> *"Their overriding conviction that the major problems facing mankind are of such complexity and are so interrelated that traditional institutions and policies are no longer able to cope with them, nor even to come to grips with their content."*

What came out of this meeting was the book, "Limits to Growth." In the book, a group of systems analysts at MIT, led by professor Jay Forrester, used the power of computers to build a computer program that analyzed five fundamental variables in numerous ways to produce a graphical representation of what the future would bring based upon assumptions built into the variables. Figure 4.1 is an example of one of the graphs in the book.[ii]

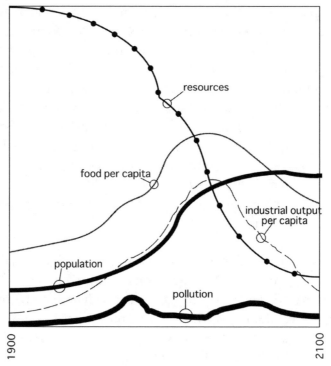

Figure 4.1: Graph Indicating Probable Future Based Upon Input Variables

The variables that they used were thought to be key to determining the point at which growth in human civilization would be limited:

Population
Agricultural Production
Resources
Industrial Production
Pollution

To people living in the early 21st century, this is like a global game of "The Sims" (a computer game with a simulated civilization), but with results that impact the future of civilization in the real world rather than on our home pc's (something that did not exist when, "The Limits to Growth," was written). The results of their computer studies were so dire, and the response to the publication so great, that a movement was born. This movement, when coupled with the then emerging environmental movement, gained influential adherents around the globe.

In the end, the result of their various scenarios all ended with the global collapse of civilization based upon exceeding one, many, or all of the five variables above. The principle variable studied was population, because the other four variables are dependent upon the size of the global population. In the scenarios modeled by this group in the early 70s, none of the demographic trends leading to decreased growth that have since begun to be revealed in population data were in evidence. In their analysis, their models were based on what is called an "exponential function" that basically means that population would accelerate until some natural limit was reached, leading to a collapse due to the overuse of resources demanded by civilization.

Based on their projections of population growth (based on the growth rate in ~1971), they projected when various resources would be exhausted. It is instructive to list those resources, and their projections in order to begin to understand the limitations of these growth models. Table 4.1 illustrates the MIT analyst's list of resources and estimated depletion dates:

Metal	Reserves	Years to Depletion (1971)	
		Static	Exponential
Aluminum	1 x 17 X 10-9 tons	100	31
Chromium	7.75 x 10-8 tons	420	95
Coal	5 x 10-12 tons	2300	111
Cobalt	4.8 x 10-9 lbs	110	60
Copper	308 x 10-6 tons	36	21
Gold	353 x 10-6 troy oz	11	9
Iron	1 x 10-11 tons	240	93
Lead	91 x 10-6 tons	26	21
Manganese	8 x 10-8 tons	97	46
Mercury	3.34 x 10-6 flasks	13	13
Molybdenum	10.8 x 10-9 lbs	79	34
Natural Gas	1.14 x 10-15 cu ft	38	22
Nickel	147 x 10-9 lbs	150	53
Petroleum	455 x 10-9 bbls	31	20
Platinum Group Metals	429 x 10-6 troy oz	130	47
Silver	5.5 x 10-9 troy oz	16	13
Tin	4.3 x 10-6 lg tons	17	15
Tungsten	2.9 x 10-9 lbs	40	28
Zinc	123 x 10-6 tons	23	18

Table 4.1: *"The Limits to Growth,"* Estimated Reserves & Resource Depletion Dates

Out of 19 resources listed in Table 4.1, the group estimated that 7 of them would be completely depleted by the year that this book is being written (2004). The "static" limits above were based on their conservative model that assumed no growth in consumption. Their exponential growth curves show 12 out of the 19 resources would be depleted by the year 2005. The fact that these resources have not been depleted is a testament to industrial creativity in finding substitutes or transcending the limits estimated in 1971 as well as the difficulty of modeling something as complex as an entire civilization.

The estimates for population made at that time have also largely not come to pass. While the population of the earth has surged from the 3.7 billion in 1970 to over 6 billion today, the rate of growth has declined. Not only has the rate of population growth declined, but the latest estimate of population in the year 2050 is 9.1 billion people, when the rate of growth will have fallen to 0.42% per year from a 1970 rate of 2.01% per year.[iii]

In, "The Limits to Growth," there is a set of underlying assumptions that go into all of the models developed by the MIT group. There are three principal assumptions that are of interest to the reader. The first assumption is that the population growth *rate* will not abate, the second is that the resources of the Earth are finite, and the third is that there are no technological solutions for the first and second problem. With 33 years of hindsight, we can see that population growth is slowing and also that technology has done a tremendous amount to hold off the limits to growth. However, as we are seeing with oil, resources are finite *on the Earth*, and solutions must be found. However, "The Limits to Growth," does not postulate any solutions based on technology. Indeed there is a deep suspicion of technology as a possible solution. The writers indicate this by the following statement:[iv]

> *We have felt it necessary to dwell so long on an analysis of technology here because we have found that technological optimism is the most common and the most dangerous reaction to our findings from the world model. Technology can relieve the symptoms of a problem without affecting the underlying causes.* **Faith in technology** *as the ultimate solution to all problems can thus divert our attention from the most fundamental problem - the problem of growth* **in a finite** *system - and prevent us from taking effective actions to solve it.*

Note the use of terms highlighted. They make the statement that *"faith in technology."* There is a huge difference between faith (which literally means the substance of things hoped for) and an understanding of the potential of technology. New technology, when coupled with the results of scientific exploration of the solar system, allows us to postulate an integrated solution that incorporates energy and resources from space. The people who wrote this book simply did not feel that it was possible for technology or space development to solve the problems related to the five variables studied in their models. Why is this?

This idea is based on what they felt were the physical limitations of the Earth. In any closed system, resources are limited. They treated the Earth as a closed system and, with this as an underlying assumption, determined that there simply were not the resources to support a greatly increased global population (as projected in 1971) no matter what technical innovations were brought forth. This is indicated in their writings a few pages later:[v]

> *The hopes of the technological optimists center on the ability of technology to remove or extend the limits to growth of population and capital. We have shown that in the world model the application of technology to apparent problems of resource depletion or pollution or food shortage has no impact on the essential problem, which is exponential growth in a finite and complex system. Our attempts to use even **the most optimistic estimates of the benefits of technology in the model did not prevent the ultimate decline of population and industry, and in fact did not in any case postpone the collapse beyond the year 2100.***

It is obvious to any of us who look to the long term future that it is desirable that our civilization goes on past the year 2100. However as the highlighted portion of the quote above illustrates, they don't feel that technology can help us transcend the limits of Earthbound resources. However correct they may be about the ultimate limits of the Earth, nowhere in the entire book did they consider the resources *off* the Earth and the ability of humanity to access and utilize these resources.

First, before the "Limits" solution is laid out, it must be stressed that I think that those that came up with these models, derived their dire consequences, and developed a plan for solving these problems, have the best intentions and feel that their solution is the only viable one for a sustainable future for civilization. However, the consequences of their solution will condemn the human race to a future of scarcity, a draconian society, and a system that, if put into place, will be extremely hard to replace. This is understood by the modelers, but to them, the choice is either to choose limits with their attendant difficulties, or have limits chosen for us as a consequence of running out of resources with dramatic and unprecedented difficulties that this sad result would bring.

Limits to Growth Solution

In, "The Limits," scenario, the modelers call for population to be limited to a number not greater than about 8.2 billion (this assumed that laws would be passed to deliberately limit the growth of families starting in 1975). They also would have limited industrial production to the level that was achieved in 1975 (this was published a year before the Arab oil embargo, of 1973). Basically, they called for a switch from growth-oriented policies that had been in place since the beginning of the industrial revolution, to a state of global equilibrium. The three primary controls in their equilibrium state were:[vi]

> *The capital, plant, and population are constant in size. The birth rate equals*

the death rate and the capital investment rate equals the depreciation rate.

All input and output rates—births, deaths, investment, and depreciation—are kept to a minimum.

The levels of capital and population and the ratio of the two are set in accordance with the values of society. They may be deliberately revised and slowly adjusted as the advance of technology creates new options.

Their idea is that the developing nations of the world must stop their growth immediately (1975) and that an intensive program of development be undertaken in the third world to bring those areas of the world to a higher state of development. After this was accomplished, growth would be tightly controlled on a global basis within a ceiling of a per capita income level of about 1/4 the 1970 U.S. level.[vii] This would require a reduction of per capita income for the United States and other advanced nations to match that level. The modelers understood the difficulty involved in such a transition and they had the following comment on that issue:[viii]

All the evidence available to us, however, suggest that of the three of the alternatives-unrestricted growth, a self-imposed limitation of growth, or a nature imposed limitation to growth-only the last two are actually possible. Accepting the nature-imposed limits to growth requires no more effort that letting things take their course and waiting to see what will happen. The most probable result of that decision, as we have tried to show here, will be an **uncontrollable decrease in population and capital…..**

….Certainly **whatever fraction of the human population remained at the end of the process** *would have very little left with which to build a new society in any form that we can now envision.*

Achieving a self-imposed limitation to growth would require much effort. It would involve learning to do many things in new ways. It would tax the ingenuity, the flexibility, and the self-discipline of the human race. Bringing a deliberate, controlled end to growth is a tremendous challenge, not easily met...

The people of the Club of Rome were and are intelligent and compassionate people. They truly feel that this is the only way to save a significant portion of humanity and build a world that is sustainable into the future. However, they make a common mistake; a mistake made by so many intelligent people throughout history. The mistake is that of discounting the future from what the advances that technology coupled with conservation can bring. Their position is espoused with the greatest sincerity and with a heartfelt faith that they are right. However, in all of their works, their models, their designs, they left out both the physical resources available from the Moon and the other worlds of our solar system as well as the technological advancements that the development of the space frontier will bring.

This was brought home to me personally by one of my mentor's, Dr. David Webb, who was on the board of directors of the organization that funded much of the MIT group's work. He related to me that he asked why the modelers did not include resources from space in any of their models. They responded that they did not think that there were any and that, anyway, they could not be used on Earth even if there were (remember that, "The Limits to Growth," was written during the Apollo Moon landings). Dr. Webb said that he was so shocked by what he felt was a cavalier dismissal of space that he vowed to dedicate the rest of his life to helping educate people about the value of space. This he has done as a founder of space studies graduate programs at the University of North Dakota and at Embry Riddle Aeronautical University in Florida as well as being a founding board member of the International Space University in Strasbourg France.

The resources of space, particularly in the near term from the Moon, are critically important to our future. These resources will help to transcend the limits to growth and help to provide a world where everyone can live at whatever level of material comfort that they care to achieve. This forms the basis of the, "Why return to the Moon," question asked at the beginning of this book. I am astonished that in the space advocacy community there has not been a more forceful linkage between space and transcending the limits to growth. This is the most powerful argument that we have, but for some reason, it is having no impression on those who write books like, "The Limits to Growth," and its successors. This linkage must be made for it should be the dominant reason for the U.S. space program. The exploration and development of the space frontier and a return to the Moon is so much more than science, it is about the future of our civilization.

There were thinkers and writers, even before and during the Apollo landings that talked about the Apollo program as a mere first step out into the vastness of space. Most of them were science fiction writers and so maybe were ignored for that reason. Science fiction was always the lesser regarded art form of that era, even Star Trek, which was a motivator of the space generation perhaps as much as Apollo, was not well enough regarded by the ratings services to keep on the air. It was only later, when the ratings services realized the inadequacy of their own methods, was it brought powerfully home that television programs concerning space as the future of our civilization were popular, leading to the almost 40 year lifespan of the Star Trek franchise.

This will be brought out as we move forward, but we need to continue to investigate the opposing solution: the solution of limits, because it is a powerful force in the global political dynamic today. Many well-meaning and intelligent people read, "The Limits to Growth," and were convinced of its warnings, including former Vice President, Albert Gore. Next we visit the book, "Beyond the Limits," by Donella and

Dennis Meadows, two of the writers of, "The Limits to Growth." "Beyond the Limits," is basically, "The Limits to Growth," updated by 20 years with their thoughts and warnings.

Beyond the Limits, Confronting Global Collapse and Envisioning a Sustainable Future

As the title of the sequel readily informs us, the writers of, "Beyond the Limits," feel that, well…we are beyond the limits of what our planet can sustain in terms of population resources, and other limits to growth. This is made obvious in their preface:[ix]

> *The human world is beyond its limits. The present way of doing things is unsustainable. The future, to be viable at all, must be one of drawing back, easing down, healing. Poverty cannot be ended by indefinite material growth; it will have to be addressed while the material human economy contracts. Like everyone else, we didn't want to come to these conclusions.*

They continue in this book to treat the Earth as a limited, closed system with no consideration of external forces, either from space resources, or from long-term threats to the equilibrium that they hope to establish. In, "Beyond the Limits," they continue to place faith in their lack of faith concerning the ability of technological progress to help us transcend the natural limits to growth on the Earth. They state this in the following way:[x]

Even with the most effective technologies and the greatest economic resilience we can believe possible, if those are the only changes, the model generates scenarios of collapse.

They are still trapped within their own limitations, never considering how resources and technologies developed off-planet can contribute to transcending the limits to growth that they so eloquently speak of. It is interesting to note that in the intervening 20 years they do gain an appreciation of how the advance in technology has made a dramatic impact on many of the problems outlined in the first book. They do acknowledge that in many areas technology has provided solutions for the substitution of materials in automobiles and in many other manufactured goods.

They do acknowledge that the advent of the microchip (the first microprocessor that forms the basis of the PC industry was first invented the year they wrote the first book and the internet had a total of less than ten computers on it, developed to survive a nuclear war doomsday scenario) enabled dramatic improvements in the efficiency of many industrial processes. The book accurately outlines the improvements in telecommunications, computers, and the fuel efficiency of cars that has helped to move the limits of the first book further into the future. However, as is illustrated in the quote above that they strongly feel that what we have done is not enough and that indeed it is impossible to do enough.

They do acknowledge that some progress has been made even on the population front, the area of most concern in 1971. However, when this book was written they felt that the growth rate was still far too high. It is ironic that, by far, most of the growth in population is a result of compassionate policies by the governments of the world to stamp out disease in the developing countries. While birth rates are declining world wide, what has changed is a dramatic decline in the death rate. In the developed world, the birth rates have dropped to that very near replacement and in the next few years some countries such as Germany, Japan, and other European countries population growth will fall below that of replacement, resulting in *population drops*. Indeed in an article on expatica.com on April 5, 2004, it was stated that the German population has already crossed this threshold and had a reduction of 143,000 in 2003 on top of a reduction of 120,000 in 2002. The German Federal Statistics Office estimates that the population of that country will fall to 75 million in 2050 from the current 82.5 million people.[xi] This trend has accelerated since 1992 when, "Beyond the Limits," was written although they recognized that this "demographic transition" was on the horizon in the developed world.

The best estimates of their modeling, a model called World3, that takes into account the effects of the demographic transition in their model run of, *infinity in, infinity out*, which estimated that, due to the forces that were starting to give rise to the demographic transition (a level of affluence comparable to the developed world plus a few decades), population would level off at somewhere around 13 billion people.[xii] The interesting thing about this model is that this is the first one that they ran that really investigated the effect of an "affluent world" scenario. However, they did not feel that this was possible due to the fact that there would not be anywhere near enough resources and that before the demographic transition took place, collapse would occur. Ah ha! Remember this piece of data because it is crucial to the position that resources from the Moon and the rest of the solar system can bring this about without resulting in a collapse. It is interesting to note that even the writers of, "Beyond the Limits," began to see that continued growth promotes population growth restraint when they say *"Industrial growth does not guarantee improvements in actual human welfare or reductions in the growth rate of population. But it certainly can help."*[xiii]

In the 12 years since, "Beyond the Limits," was written, there has been a continual improvement in the population growth picture overall. In a table titled "Additions to World Population, 1971, 1991, the beginnings of the decline in population growth was noted.[xiv] Table 4.2 presents their data, plus data from the U.S. Census Bureau:

Year	Population (Millions)	Growth Rate (Per Year)	People Added (Millions)
1971	3600	2.1%	76
1991	5400	1.7% (1.55% revised)	82 (84 revised)
2004	6375	1.14%	73
2050	9084	0.42%	38

Table 4.2: Population Estimates from "Beyond the Limits" and U.S. Census Bureau

The interesting aspect of the data presented above is that even in absolute terms population growth is decreasing. The supplemental data above is the latest revision to the U.S. Census Bureau's international database from 2003.[xv] In the latest data from both the Census Bureau and from the U.N., global population is expected to top out at about 9.1 billion people around the year 2050. This is a remarkable turn around from the 13 billion estimate from the World3 model in 1992's, "Beyond the Limits." What is the reason for this demographic shift? To put it simply: economic growth. While there has been some improvement in family planning, the world has also experienced a lot of economic growth over the last 12 years, even with the 1998 Asian recession and the dot com bust in the United States and Europe. The world has been literally transformed by the advent of the global Internet, something just invented in 1971 and just beginning to spill into the public domain in 1992. Much of this growth has come without any per capita increase in many of the key measures of resource growth in the U.S. However, in some nations there has been a negative growth rate for several years.

In Japan and Western Europe, the decline in growth rate to near replacement can easily be attributed to affluence. This is not the case in Eastern Europe and in some areas of Africa where the AIDS epidemic has been the cause for the death's of over 20 million people in the last twenty years with the prospect of 40 million more deaths over the next 10 to 15 years. Indeed in some of the wilder conspiracy theories rampant in Africa, AIDS is the dark side of the efforts by international agencies to control the population growth demanded by the more strident adherents to, "The Limits to Growth," scenario. While I do not for a second attribute dark motives to those who wrote Limits to Growth or its sequel, there are many in the world who do and this factor alone works against some of their goals. This is something that the modelers never considered but should have. Also beyond the AIDS problem, fighting in Rwanda in 1994 alone resulted in the extermination of 1.5 million people just for the crime of being different from their fellow countrymen.

Growth is a natural impulse. In nature growth happens until natural limits produce counter-pressures that result in the reduction in populations. In the state of Alabama where I live, for example, the deer population has grown enormously, because of the killing of their natural predators by humans. Deer hunters can see the results of this unbridled growth when the deer that they kill are lighter in weight and less healthy.

Since the only significant predators are humans, the Fish and Game Department authorizes greater kills in the overpopulation years, restoring the deer population to balance with the natural food supply. The deer grow in relative size and health, which is duly noted by the increase in business at trophy shops in deer hunting areas. The modelers in, "Beyond the Limits," understand that mankind has the growth impulse but has always counted on the ability of people to consider their future and make what they feel are rational decisions to reduce growth to equilibrium without the natural collapse that accompanies exceeding natural limits. As we see on the news every day, mankind is far from rational.

The writers of, "Limits to Growth," and, "Beyond the Limits," are intelligent, well meaning people and they do understand that moving from growth to equilibrium is exceedingly difficult. They address this difficulty thus:

> *How in practice, can anyone attack these problems?... It will require every kind of human talent. It will need not only technical and entrepreneurial innovation, but also communal, social, political, artistic, and spiritual innovation...*

In other words they are asking that we change the entire structure of human civilization from that of growth to one of equilibrium. However, they do not really develop a way to do this and they realize that the problem in doing this may well be unattainable. They are wise in this but they still have the basic fallacy that limits are inevitable and that there is nothing that technology can do to help us transcend the limits that are intrinsic to a finite world. The evidence contradicts this and the data on population developed in the intervening 12 years indicates that their primary concern, population, is going to top out at numbers well below their projections. The number is still over a billion above what their models say are possible for the Earth's current limits to support but close enough to give hope. The maximum limit in their scenarios was set at 7.7 billion people, if their policies for stopping growth had been enacted in the early 1990's.

Obviously this did not happen, so in their minds, we have gone beyond the limits. That is where the resources from the Moon and beyond can come in to help transcend these limits and bring us to a world where those 9.1 billion people can not only live, but live in a world civilization where they can feed themselves and reach a level of affluence far above what is the norm in the developing countries today or the expectations of the "Limits" modelers. This is the promise that the general exploration and development of space brings.

Before we illustrate the advantages of returning to the Moon and beyond, let us continue with the general theme of the limits to growth with a book from the man who took this idea further and made it his personal crusade that led him to within one heartbeat of the presidency in 1992 and within three electoral votes in the year 2000. It is important to review Mr. Gore's, "Earth in the Balance," because the philosophy embodied in the movement resulting from the, "Limits to Growth," and, "Beyond the Limits," are now a part of the global political mainstream. The idea of limits are at the core of global agreements such as the Kyoto Accords, a global agreement that almost came into force of law in this country as well as around the world and would have contained central aspects of the philosophy of moving from growth to equilibrium. The administration that Gore was a part of, was very much in favor of this agreement, so we need to look closer at what Mr. Gore wrote to see the connection.

Earth in the Balance

In 1992, Senator (soon to be Vice President) Albert Gore wrote his variation of, "The Limits to Growth," called, "Earth in the Balance, Ecology and the Human Spirit."

This book took a slightly different slant on the issues enumerated in, "Limits," and, "Beyond the Limits." Mr. Gore uses the issue of "Global Warming," associated with the increasing concentration of CO_2 in the atmosphere that is leading us to global calamity, as his call for action. Mr. Gore does have a lot more familiarity with some of the more specific subjects that he talks about related to global warming than the more dry general treatment of the Meadows' (the Author's of Limits to Growth). He mixes this with the more general theme of population growth (which is partly the cause of the CO_2 problem), as well as nuclear weapons, to develop a theme that equates industrial growth, climate change and population problems to the danger of nuclear war. For example:[xvi]

> *Like the population explosion, the scientific and technological revolution began to pick up speed slowly during the eighteenth century. And this ongoing revolution has also suddenly accelerated exponentially. For example, it is now an axiom in many fields of science that more new and important discoveries have taken place in the last ten years than in the entire previous history of science. While no single discovery has had the kind of effect on our relationship to the earth that nuclear weapons have had on our relationship to warfare, it is nevertheless true that taken together, they have completely transformed our cumulative ability to exploit the earth for sustenance— making the consequences of unrestrained exploitation every bit as unthinkable as the consequences of un-restrained nuclear war.*

While Mr. Gore may exaggerate concerning the pace of technological advance, he does share with the Meadows the common theme that the industrial revolution is where the problem began, that technology is not the answer, we are already beyond the limits, and that radical change in human social organization is necessary. Here are some of the quotes that illustrate the similarities:[xvii]

Our challenge is to accelerate the needed change in thinking about our relationship to the environment in order to shift the pattern of our civilization to a new equilibrium— before the world's ecological system loses its current one.

Note the similarities with, "Beyond the Limits," concerning the shift to equilibrium from growth. However, Gore does make a stronger tie directly to the environment than Meadows does. Gore does make good use of history by expressing the impact of very recent climate change on society. However, he makes several mistakes in his treatise in terms of the numbers that he uses (to mention one, his statement that global temperatures are the highest in 12,000 years is wrong)[xviii], but in general tone he illustrates some of what he feels are concrete examples of the general trends that, "Limits to Growth," and, "Beyond the Limits," treats in a more analytical fashion.

One obvious and vast difference between the Meadows' and Gore is that Mr. Gore had a major influence on government policy in his Senate position and then in the White House itself for eight years in the 1990s. Therefore Mr. Gore should have a wider view than that espoused in the "Limits" books. However, he uses the same language as the Meadows' but in a much more strident way such as:[xix]

We have also fallen victim to a kind of technological hubris, which tempts us to believe that our new powers may be unlimited. We dare to imagine that we will find technological solutions for every technologically induced problem. It is as if civilization stands in awe of its own technological prowess, entranced by the wondrous and unfamiliar power it never dreamed would be accessible to mortal man. In a modern version of the Greek myth, our hubris tempts us to appropriate for ourselves—not from the gods but from science and technology—awesome powers and to demand from nature godlike privileges to indulge our Olympian appetite for more.

Perhaps as a politician, Mr. Gore feels the necessity to be more theatrical in his language in order to get the point across, but some of the analogies that he uses are interesting in their extreme imagery. He goes so far as to equate not taking action to move toward global economic equilibrium to an "ecological Kristallnacht," which has to be read to get a sense of the urgency that he places upon the problems that Limits to Growth simply states:[xx]

Now warnings of a different sort signal an environmental holocaust without precedent. But where is the moral alertness that might make us more sensitive to the new pattern of environmental change? Once again, world leaders waffle, hoping the danger will dissipate. Yet today the evidence of an ecological Kristallnacht is as clear as the sound of glass shattering in Berlin....We have a clear choice: we can either wait for change to be imposed on us—and so increase the risk of catastrophe—or we can make some difficult changes on our own terms, and so reclaim control of our destiny.

He goes on in his book to equate the current state of global denial to that of alcoholics, a dysfunctional family, and the evil of corporate greed. Mr. Gore's rhetorical hubris not withstanding, his underlying assumptions and his proposed solution set is remarkably similar to that of, "Limits to Growth," and, "Beyond the Limits." Also, explicitly laid out is his common dismissal of technology as an answer as he states:[xxi]

It is important, however, to remember that there is a great danger in seeing technology alone as the answer to the environmental crisis. In fact, the idea that new technology is the solution to all our problems is a central part of the faulty way of thinking that created the crisis in the first place.

This is a common blindness that would not be as bad if this had not come from the second most powerful executive in the nation. The danger in the shift from the more academic treatment of the Meadows' to the, "Earth in the Balance," stridency of Gore, is that his proposed solutions either were implemented as national policy or have been given increasing credibility. An example of this is his proposed policy on population control. He does recognize the existence of the significant "demographic transition" but mistakes the reasons for it. (As an aside at this moment I urge the reader to do an Internet search on the term "demographic transition" and do research that is beyond the scope of what can be examined in this book). His first strategic goal proposed was:[xxii]

The first strategic goal should be stabilizing of world population, with policies designed to create in every nation of the world the conditions necessary for the so-called demographic transition.

Mr. Gore even goes as far as to recognize that the growth of per-capita income is central to the demographic transition but declares that this change would take centuries. As the data indicate in Table 4.2 this transition will occur as soon as 2050 and could, based on current trends, happen sooner. This is the mistake common to Gore and the Meadows' in thinking that it will take extraordinary moral changes in human nature in order to reach the goal of sustainability. No one has ever gotten elected, or rich, betting against human nature and there is absolutely nothing but their faith in the doomsday scenarios that they paint and its effect on the global public, that would lead one to believe that these changes will happen. Using human nature as a guide, it would seem that growth is key to bringing about the demographic transition, especially in the developing world, however, Mr. Gore's final solution is:

Finally, the plan should have as its more general, integrating goal the establishment, especially in the developing world—of social and political conditions most conductive to the emergence of sustainable societies—such as social justice (including equitable patterns of land ownership); a commitment to human rights; adequate nutrition, healthcare, and shelter; high literacy rates; and greater political freedom, participation, and accountability.

The above illustrates again what seems on the surface to be a very laudable goal, but one that in the real world will take longer to achieve than the horizon for global collapse. The evidence for this is clear and is in the news every day in looking at the violence in Iraq, Israel, and the rest of the Middle East and Africa. The U.N. could not even stop the slaughter in Rwanda in 1994 or in Sudan today (conspiracy theorists use the Meadows' and Gore as illustrations that the U.N. does not want to stop these tragedies as a convenient way to reduce population). While the conspiracy theorists can be dismissed by those of us in the developed world, this type of thinking dominates many parts of the globe and acts as a counter-pressure to the types of changes that Meadows and Gore say are absolutely necessary to build a stable world. How would this conflict be resolved? If history is any guide, it will be in blood. Would it not be better if this could be avoided and the centuries that it may take for common sense to take hold in some parts of the world to be made possible? Again, this is what the resources of the Moon first, then the rest of the solar system will bring to our civilization.

Solutions?

In this chapter, I have explored what is one of the dominant political movements today in terms of where many would like our civilization to go in the future. It is clear that any proposal to shift from what has been centuries of growth to one of global limits is one that would be difficult in the extreme to achieve. This would require not only that the developing world be brought upward in standard of living (this is explicit in the United Nations Declaration of the Right of Development General Assembly resolution 41/128 of 4 December 1986)[xxiii], as well as the reduction of the developed world's per capita standard of living by a factor of 4 to 10. This simply is not going to happen unless there is no other choice. This is the fundamental flaw of, "Limits to Growth," its sequel, and, "Earth in the Balance." Both the Meadows' and Gore say that we don't have a choice. I, and a great majority of my peers in the space arena, say that we do.

All you have to do to look for evidence that there is at least the possibility of resources from space is to look up at night when the sky is clear and the Moon is visible. Look at its pockmarked surface. That is the result of millions upon millions of asteroids slamming into it over the history of the solar system. Many of these asteroids are rich in water, hence the recent discovery of water in the permanently shadowed regions at the poles of the Moon. Mars has been shown to have literally oceans of water trapped beneath its surface. On the Moon as well, a significant percentage of these impacting asteroids have rich deposits of metals that can be used to establish the hydrogen economy as well as enable a significant off world presence by mankind. Some of the *smaller* metal asteroids are so rich in resources that they exceed the total amount of metal mined in the history of mankind!

This is where we are going in the next chapter on the resources available and how they integrate into the earthbound economy to help us transcend the limits to growth and enable a planetary civilization to live at a level that would give the poorest among us comforts unknown in all of history. Even beyond this, we have a real chance and a real responsibility to do this for our future. What the Meadows' ignored and Gore examined extensively, was that we live in an uncertain world. Just one volcanic eruption in 1816 led to three years of famine in Europe and gave full birth to the Luddite movement in Britain. Within the recorded history of mankind, the Sahara was grassland, and we have lived through the Little Ice Age. Go back a little further to the end of the last Ice Age, 12,000 years ago, and you see the Oceans rising over 430 feet! This will be looked at later, but the point that needs to be made is that, no matter what humans do or do not do, external natural factors will make life difficult for all of us, it is only a matter of time.

*　　　　　*　　　　　*　　　　　*　　　　　*

[i] D. H. Meadows, et al, *The Limits to Growth*, Potomac Books, 1972 p. ix

[ii] Ibid: Figure 48 p. 173

[iii] http://www.census.gov/ipc/www/worldpop.html

[iv] Limits to Growth, p. 139

[v] Ibid, p. 152

[vi] Ibid, p. 179

[vii] Ibid, p. 169

[viii] Ibid, p. 174-175

[ix] D. H. Meadows, et al, *Beyond the Limits*, Chelsea Green Publishing Co, 1992, Post Mills Vermont, p. xv

[x] Ibid, p. 162

[xi] http://www.expatica.com/source/site_article.asp?subchannel_id=26&story_id=6334

[xii] *Beyond the Limits*, p. 120

[xiii] Ibid, p. 32

[xiv] Ibid, p. 24

[xv] U.S. Bureau of the Census, International Data Base. Note: Data updated 7-17-2003

[xvi] Earth in the Balance, Al Gore, Houghton Mifflin, New York, New York, 1992, p. 31

[xvii] Ibid, p. 48

[xviii] Ibid, p. 105

[xix] Ibid, p. 207

[xx] Ibid, p. 177

[xxi] Ibid, p. 317

[xxii] Ibid, p. 305

[xxiii] http://www.unhchr.ch/html/menu3/b/74.htm

Chapter 5:

Our Energy Future and its Impact on Civilization

Plentiful and inexpensive energy has been the key to our civilization over the past 300 years and will continue to be so in the future. In a previous chapter it was pointed out that our use of energy, from coal to oil, has had a direct impact on dramatically increasing our life expectancy and for some, the quality of that longer life span. It was also pointed out that our primary energy source today, oil the magic elixir that has enabled a planetary civilization, is declining as a resource and it is only a matter of time, measured in decades, before we will be beyond the peak of its availability and on the way to its eventual exhaustion. In the chapter on the "Limits to Growth," a very dire picture was painted of a general depletion of most of the primary resources that make our civilization function. It was further pointed out that the authors of these books believe that with existing means there is no way to bring the rest of the world to our level of affluence without breaking the limits to growth, leading to a catastrophe of plunging populations, exhausted resources, war, and little ability for us to cope with what is left.

It was also previously noted the solutions of those who have faith in the "Limits to Growth," suggested that in order for civilization to be saved, dramatic sacrifices would have to be made, and that a population of the size that we know will exist by 2050 cannot be sustained by the resources of the Earth. More disturbingly, the authors have developed, as a central facet of their underlying assumptions, the position that technology cannot help get us extend or transcend the limits that they so eloquently present. Mr. Gore extends this thought pattern with his holocaust analogies and quasi-religious metaphors. However, is this the inevitable path that humanity and civilization has to take? Are we doomed to a world of increasing scarcity, limited opportunity, and at a level that most in the developed would consider impoverished?

The answer is an unvarnished, unequivocal, unqualified no! This is not a case of "blind faith" in the ability of some amorphous historical force called technology to solve our woes. It is based upon very real knowledge of energy sources on the Earth and in space that are orders of magnitude greater than we use today. It is based upon very real evidence of reserves of metals, minerals, and the other building blocks of civilization that we know exist on the Moon, in asteroids and in free space. Since the original "Limits to Growth" and other books of like mind were written our technology has advanced an incredible amount.

When Mr. Gore penned "Earth in the Balance," the Internet had not yet exploded into the general population's everyday lives. That advance alone has completely changed the way that civilization is organized with huge increases in productivity and efficiency in the global distribution network. After rising for hundreds of years, beginning after the Arab oil embargo of 1973, per capita energy use in the U.S. has remained relatively constant, even with the explosion of new electronic devices in our homes and office. Advances in technology have accomplished this, from the "Energy Star" rating on household appliances, to heat pumps, to automobile fuel economy. It is the advance of technology that has allowed us to beat the estimates that 12 out of 19 critical materials listed in Limits to Growth as depleted by 2005.

There is a blindness that afflicts those who reject new ideas and the possibility of technological advances to solve civilization-wide societal problems. Before a technical answer is even considered it is rejected because of a faith that technology does not hold the answers. The danger in this thought pattern is that when obvious technical solutions do present themselves they are rejected, not because of any lack of intrinsic merit but because their faith blinds them to solutions that are technical in nature. The solutions that space and technology advocates talk about are real, they are known, and they can be implemented for reasonable expenditures coupled with incentive laws that enable private enterprise to contribute its part to the solution.

Technology and capitalism are not genies that must be stuffed back into their bottle in order for us to survive, they are tools to enable all of us to transcend poverty and the limits to growth that a single planet imposes upon all of humanity. We have a choice, the future of Mel Gibson's "Mad Max," and its dismal counterparts, or the future of Captain Kirk and Picard's "Star Trek." I know the one that I prefer, and so will the majority of people if the decision is placed before them. To get there we must develop new resources of energy, coupled with resources and the ingenuity of the human race that has transcended every limit to growth that has ever been placed before us.

Energy and Resources of the Earth Today

Solar Energy

It is a scientific fact, that at the top of the Earth's atmosphere in space, an average of 1,358 watts of solar energy fills every square meter of space. At the Earth's surface this is reduced to about a thousand watts per square meter, still a formidable number. The average American style house rarely uses more than 10 kilowatts of power even at peak demand. If there were some way to convert all of the energy of the sun into electrical power it would take about 10 square meters of area during the day, and for a good margin to make up for the night it would take about 40 square meters. This amount of area makes up for cloudy days as well. This is an area about 13 feet by 40 feet in area, which is smaller than the roof of a modest two bed room house. In total, the Earth receives every day a total of 18,000 terawatts of power! The total amount of energy used today globally is no more than 9 terawatts. If every single person on the earth in 2050 was brought to the same level of affluence as is the norm for the affluent in the United States that would require no more than 90 terawatts of power. While that is ten times the amount that we use today it is less than one half of one percent of the solar power that hits the Earth. There is obviously not a shortage of solar energy, just a shortage in current technology of how to capture that energy for our use.

Unfortunately it is impossible to capture all of the solar energy that falls on the Earth and convert it to power. Today the best solar cells that you can purchase are made for satellites and convert about 28.5% of the energy from the sun into DC electrical power. That means that in space a square meter (about 10% larger than a square yard) of solar array would produce 380 watts of power and on the ground about 280 watts. These cells can be purchased from Emcore or SpectroLab, two American companies in the market.[i] That is a huge increase in efficiency from the 8% that was state of the art at the time that "The Limits to Growth" was written in 1971. This is also much better than the 18% efficiency that was the norm in 1992 when "Beyond the Limits" was written. However, these are very expensive cells that rely on semiconductor production equipment to make them, meaning that today they cost about $300 per watt of power produced. This price is much lower than the equivalent price in 1992 and 1971 but still quite expensive.

There are alternatives to the high tech solar cells that are available at most high tech electronics outlets. They are made of multi-crystalline silicon and have efficiencies of between 10-15%. They only cost about $10 per watt at the retail level and are used extensively for recreational power supplies, remote power for things like roadside call boxes, warning lights, and even supplemental and primary home power. They are still more expensive than electricity from the grid but many American states have tax breaks for the installation of such systems that lower the total cost of ownership. This technology has come a long way since the 1970's and still has a way to go to become cost effective for home applications outside of desert areas.

There are some exciting new developments over the past few years that will shatter previous records in solar energy development. The most efficient solar cells that can be purchased from a company like Emcore are composed of three layers of dissimilar materials. Figure 5.1 shows how these cells are constructed:

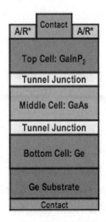

Figure 5.1: Emcore Tandem Solar Cell Construction (Picture Courtesy Emcore)

The three different materials are Germanium (Ge), Gallium Arsenide (GaAs), and Gallium Indium Phosphide (GaAsInP). Each one of these materials absorb solar energy at a different set of wavelengths and convert it to electrical power. In Quantum Mechanics this is called the "bandgap energy". Each photon of sunlight has a specific amount of energy based upon its wavelength. These different wavelengths can be seen in the various colors of a rainbow with red having the lowest energy (longest wavelength) and violet having the highest energy (shortest wavelength). A "multi-bandgap" solar cell converts more than one color "band" of energy into electrical power. This is based upon physically stacking the three different types of materials on top of each other and allowing the solar energy to pass through each one to absorb their preferred wavelengths.

The exciting new development in solar cell technology is the "Quantum Dot." These cells are made out of a single crystal of semiconductor made from Copper (Cu) Indium (In) and Selenium (Se) grown as one crystal with the formula varying at different points in the crystal. This is what is called a homogenous crystal structure (a single crystal) with different levels of impurities purposely put in to make the dot work. This converts an even broader band of solar energy with a theoretical maximum of 63.5% possible![ii] In comparison, today's multi-bandgap cells, such as shown in figure 5.1, are called heterogeneous structures, which means different types of materials physically bonded together. The Quantum Dot cell should be much cheaper to make since it is made in one process rather than many cells made separately and physically stacked onto each other.

Dr. Nevelle Marzwell, a researcher at the NASA Jet Propulsion Laboratory, and his team assembled such a cell and then added a thermal cell to that to convert the waste heat from the cell into electricity. Dr. Marzwell reports that they reached a demonstrated efficiency of 63% with these cells in the lab.[iii] Unfortunately, after supporting this research effort for four years the Congress refused to fund any further development efforts and this incredible technology sits in the lab waiting for funding to bring it to market. An example of the potential of this technology is that the entire 10 kilowatt power of an affluent American home could be provided by a solar array of this material that is only about 25 square meters for homes in most of the country.

In the United States, as well as in most of the world, it is coal that powers the majority of centralized power plants. Coal that produces soot and other pollutants as well as vast quantities of CO_2 dominates electrical power generation, yet this advanced solar cell technology is languishing after Congress cut the funding. An organization that I was a member of, Prospace, lobbied successfully for this funding for three years that enabled the progress described above. This research was originally for a space solar power satellite that would beam power to the ground and the Quantum Dot is an incredible spin off of that modest effort that still merits funding. Why is it not funded now? The Quantum Dot is not funded due to an incredible lack of foresight and understanding by many of our legislators and the venture capital community who ought to know better.

These types of solar cells would enable more powerful spacecraft in space as well as profitable solar power production on the Earth. After solar arrays are built they emit zero CO_2 while providing huge amounts of power. It is interesting to note that over the past 20 years almost all advances in solar cell technology has been funded by NASA or the Department of Defense for space applications. This also reveals the blindness of some in the environmental movement who have bought into the "technology provides no solutions" side of the argument. The more environmental minded persons in Congress should be the champions of such technology but because it has been traditionally supported by the military and NASA it is automatically to be opposed by some.

Solar energy could provide significant quantities of power in areas of the world that are in most dire need of power without the huge recurring costs and foreign exchange losses associated with oil or gas fired plants. Solar power systems are more productive in areas close to the equator or in desert areas of the world. Solar energy can run pumps to bring water for people and agriculture in Africa and South East Asia. Solar energy can provide power for telecommunications so that people in remote areas of the world can become part of the world community through the Internet, and bring many other benefits to those in the developing world. This can be done without the continual bleeding of scarce cash for oil.

Solar power will probably never be extensively used in the high latitudes of Europe, Asia, and North America due to cloud cover, low angles to the Sun, and short daylight periods during winter. With these limitations in mind, the development of advanced solar cell technology like the Quantum Dot can really benefit the developing world on Earth as well as bring significant benefits to satellites and power systems based on the Moon or Mars. Space Solar Power satellites are one speculative power system to provide power for the Earth that would greatly benefit from these advanced cells as well. This subject will be returned to later.

Nuclear Power (Fission)

Nuclear Fission power has been a part of our lives for over 40 years. According to the Bulletin of the Atomic Scientists there are 429 fission reactors operating around the world. These reactors produce a staggering 345 gigawatts of electrical power, about seven percent of total planetary electrical demand. Oil, coal, and natural gas fire the turbines that provide most of the rest of the global power supply with hydroelectric and other renewables filling the gap.[iv]

As of 2003 the U.S. has 103 licensed reactors generating 20% of the total electrical power demand for the nation. Nuclear fission reactors generate a higher percentage of power than any other source except for coal in the U.S., which has more reactors than any other country. In 2002 nuclear fission reactors generated more than 772 billion kilowatt hours worth of electrical power, more than the entire national electrical power output of the United States in 1963.[v] Even though no nuclear reactors have been licensed since 1978, electrical power generation by fission reactors has increased by one third between 1990 and 2002. This was due to increased productivity at existing plants, the addition of five new reactors licensed prior to 1978, and modifications to existing systems.[vi] Around the world, Japan generates 28% of its power, Belgium generates 60% of its power, France generates 78% of its power and Lithuania generates a total of 82% using obsolete Soviet designed Chernobyl type reactors.

These statistics belie the problems that beset Nuclear Fission power. The two problems of radioactive waste and the potential for the spent nuclear fuel rods to be reprocessed into a nuclear weapon limits the utility and the desirability of expanding nuclear fission power dramatically in order to take over more load from oil, gas, and coal, in generating power and reducing emissions of CO_2 and other pollutants. Even with these problems sources indicate that there are 30 plants under construction in Asia with a generating capacity of about 22 gigawatts.

In the United States a few plants, such as Browns Ferry Unit 1, mothballed for 20 years, are being restarted, and other plants, such as the nearly completed Bellafonte, in northeast Alabama are being considered for completion. Data compiled by the Nuclear Energy Institute list 97 power plants cancelled that were either under construction or licensed to begin construction in 1978 representing over 107 gigawatts of power.[vii]

This amount of power would have boosted the fraction of power generated by these reactors to about 28% of total U.S. electricity production. This increase would have meant that millions of tons of CO_2 that are currently pumped into the atmosphere from oil and coal fired plants, would have not occurred, making a significant impact on our total national CO_2 pollution mitigation efforts. This is an example of how environmentalists are also governed by the law of unintended consequences. I remember bumper stickers in the 1970's that proclaimed "Split Wood, Not Atoms." Today, we know that this adds to the burden of CO2 build up as well as removing the trees that sequester large amounts of that greenhouse gas. However, even with this in mind, the intractable problems of nuclear waste and proliferation of nuclear weapons, means that fission has a limited future in power generation on the earth.

Nuclear Fission in Space

During the 1950's and 1960s an ambitious program to develop a nuclear engine for space was carried out by the U.S. Atomic Energy Commission (AEC) and the U.S. Air Force. Later NASA took over the program and incorporated this technology into the Apollo program. According to testimony to congress in 1966 this engine could increase the payload delivered to the Lunar surface from 28,000 lbs to 45,000 lbs and could place a payload into Mars orbit of about 50,000 lbs.[viii] Figure 5.2 gives a picture of the NERVA engine that would have powered Apollo 20 to the Moon.

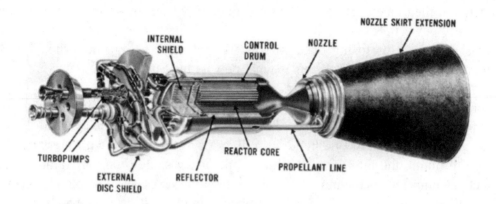

Figure 5.2: NERVA 75,000 lb Thrust Engine (Picture Courtesy NASA)

What is amazing is that a ground demonstration version of this engine was fully tested and the flight engine built. This engine and its associated upper stage (a modified Apollo upper stage) could have been refueled in orbit for multiple trips between Low Earth Orbit (LEO) and lunar orbit. With hydrogen fuel delivered into earth orbit by tankers, an earth-to-moon round trip transportation system would have been put into place that would have enabled the development of a lunar economy thirty years ago. This technology was brought to near fruition to support a robust space exploration and development program. However, the NERVA engine was cancelled in the general malaise that gripped the United States in the early 70's and the Apollo 20 launch vehicle is a billion dollar lawn ornament at NASA in Houston. Today parts of the NERVA sit at the U.S. Space and Rocket Center in Huntsville Alabama, a further testament to the fickle nature of government directed space programs.

Recently, with the advent of the O'Keefe administration at NASA, interest in nuclear fission engines for space applications has been revived. These engines, while very capable, face opposition by those to whom the very word nuclear is cause for violent opposition. They would provide an incredible increase in capability to move large payloads around in space and it may be that a valuable part of a Lunar economy would entail the construction of such engines and vehicles there as president Bush intimated in his 2004 speech. Flying from the Moon to the asteroids, Mars and beyond, they would have no chance of producing a launch accident in the atmosphere of the Earth.

Nuclear Power (Fusion)

It has been said by some that nuclear fusion is a technology whose time will always be in the future. Nuclear fission occurs when atoms of Uranium or other heavy radioactive elements are bombarded with neutrons. This splits these heavy atoms, releasing energy and more neutrons that keep the reaction going. This energy is in the form of heat, that is used to heat water or other liquids, to turn turbines and generate electrical power. Nuclear fusion occurs when atoms of hydrogen or other very lightweight atoms are slammed together with sufficient force to overcome the repulsion of their electrons and get close enough to where the strong nuclear force is sufficiently powerful to allow the atoms to "fuse" together and create heavier atoms such as Helium. This fusing of atoms releases far more energy than the fission of heavy atoms and a variation of this is what powers the Sun.

Fusion was over hyped in its early days with proponents promising inexhaustible energy supplies as early as 1980. The difficulty is that fusion is hard! It takes temperatures in excess of one hundred million degrees coupled with pressures high enough to mimic the interior of the Sun while at the same time maintaining a near perfect vacuum. This is the most difficult engineering feat ever attempted by man, far more difficult than going to the Moon. However, in the past fifteen years tremendous progress has been made and we are now in the position that we now know the majority of what we need to know in order to make fusion a reality.

The history of fusion over the past 51 years since U.S. government funded research began is one of steady progress. Unfortunately it took 25 years for the first really significant milestone to actually occur. In 1978 a research reactor achieved temperatures in excess of 58 million degrees centigrade, the minimum required for a self-sustaining fusion reaction. In 1982 the Tokamak Fusion Test Reactor (TFTR) entered service, leading to major advances based upon an original Russian design. In 1985 the Spherical Torus reactor was conceived, that allowed for a higher plasma pressure for a given magnetic field strength. In 1994 the TFTR reactor achieved an output power of 10.7 megawatts. Following on this milestone the next year the TFTR reactor set the record for highest temperature ever recorded for a laboratory plasma, more than thirty times hotter than the center of the sun![ix]

One of the amazing and hopeful things about nuclear fusion research is that it has been a global scientific endeavor that has transcended the cold war and other issues of international nation state competition and conflict. The key technological breakthrough in magnetic fusion came from the depths of the secretive Soviet atomic energy program. Other breakthroughs have come from individual scientists and teams from virtually every advanced nation in the world. The nuclear physics community has managed to maintain a collegial environment in the face of incredible odds. This atmosphere of cooperation was one of the primary factors that led to the creation of the World Wide Web or the Internet that we know today.

The web originated as a collaborative tool used by physicists at Geneva's CERN laboratory to communicate efficiently with their colleagues around the world. This was coupled with a group of students at the University of Illinois who developed the original web browser Mosaic, under a National Science Foundation grant at the National Center for Supercomputer Applications. Mark Andreeson, the founder of

Netscape, was one of those students. It is interesting in the extreme that the foundations of the dot com revolution, the global information society, and the incredible improvement in our lives that resulted are built upon the desire of scientists for efficient collaboration in nuclear fusion research.

Fusion today is at a critical crossroads. Beginning in 1990 the design of the most advanced reactor yet was begun. This reactor, called the ITER (or "the way" in Latin) is another in a long line of international collaborative efforts, is almost the final link in a long chain of research reactors preparing the way for the commercial viability of nuclear fusion. Figure 5.3 is a graphical representation of the ITER reactor.

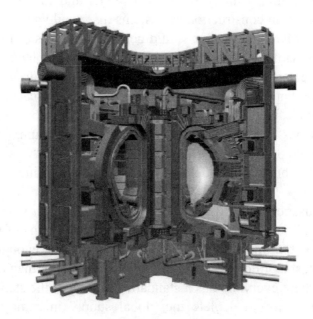

Figure 5.3: The Core of the ITER Reactor (Courtesy of www.iter.org)

This reactor was originally slated to begin construction in the late 1990's but was held up by disputes over its cost. These disputes led to the exit of the United States from the ITER program in 1998 as the U.S. was unwilling to shoulder its portion of the proposed cost. It is interesting to note that this happened while the vice president at the time was Albert Gore, the author of "Earth in the Balance." The European Union, Japan, and Russia, continued on with a de-scoped ITER design, which was approved in the year 2000. Several sites around the world have competed for the home of the ITER and in 2003 China joined the ITER as a full party, Korea joined as a limited party, and the U.S. rejoined the consortium. As of May 2004 there is an impasse over the site of the ITER between Europe and Japan that is yet to be decided. This impasse has lasted for three years now and it is hoped that it can be resolved soon.

The ITER reactor is designed to be able to generate up to half a gigawatt of power and should exceed the power input to it by up to ten times. This is called the Q factor and it relates to the amount of power that has to go into the reactor to start it and then generate an excess of power, thus sustaining the reaction. This will not be the first time that a Q greater than 1 (more power out than in) has been achieved but it will be the first time that the reaction will be sustained for long periods of times, measured in minutes. This would be an incredible milestone that will help scientists and engineers solve most of the remaining problems associated with plasma confinement, materials degradation from neutrons generated from the reactor, tritium fuel breeding, and reactor energy extraction. The answers to these questions will give the scientists and engineers the data that they need to design a prototype of a commercial reactor called DEMO, currently scheduled for development in the 2020 timeframe.[x]

The DEMO reactor would actually serve as a model and functional prototype for power companies to begin construction of systems that would generate power for the electrical grid. A DEMO class reactor would generate upwards of three gigawatts of electrical power, the same order of power as a major fission plant. The fuel for this reactor would be deuterium and tritium, both isotopes of hydrogen. Deuterium is common in sea water but tritium has to be created in a nuclear reactor. Current design ideas for commercial fusion reactors use a liquid lithium-6 blanket, that when bombarded by neutrons from the fusing of deuterium and tritium, create an atom of helium-4 and one atom of tritium.[xi] What this means is that the "waste" neutrons from the reactor are used to breed more fuel in what is called the tritium-breeding blanket.

A reactor such as this, generating 3 gigawatts of electrical power would use about 300 kilograms (660 lb) of deuterium a year and about 450 kilograms of tritium (990 lb). There is enough deuterium in the water in the oceans for enough power to run civilization for at least a billion years. Lithium is less common than deuterium and to run a civilization where everyone has a standard of living at the standard of the U.S. (around 90 terawatts) would deplete the global supply in somewhat less than a thousand years! That is the good news, there is bad news in this story.

A fusion nuclear reactor using the deuterium/tritium fuel mixture would bombard the walls of the reactor with neutrons that would transmute the metals of the wall into radioactive byproducts. Current research is showing that there are materials that can be used for the reactor walls that would lead to radioactive byproducts that would have a half life of only about one hundred years. This is in comparison to the byproducts of a fission reactor that have a half life of tens of thousands of years. Also, the good news is that the radioactive byproducts are much easier to contain (the metals in the walls and other solid materials), all in all a pretty good solution to the problem of nuclear power.

Space and Fusion

This discussion of fusion may seem to be bringing another unrelated subject into a discussion about a commercial lunar development effort but this is not the case. The problems of energy, resources, pollution and population are all interrelated and dependent upon population. The more prolific utilization of energy from coal and oil were the thing that got the population growing in the first place and if fusion is our long term solution to the energy problem then we have won half of the battle to transcend the limits to growth. This is due to the fact that populations have been demonstrated to make the "demographic transition" mentioned in the previous chapter, from rapid growth, to replacement in affluent societies.

This is a dramatic trend that was not seen in the 70's and bodes well for the future. However, fusion is also dependent upon space based resources in the long term, namely for the fuel—helium-3—which is exceedingly rare on the Earth, very rare on the Moon, and plentiful on all of the gas giant outer planets of the solar system.

The fusion reactor described above, while far superior to a fission reactor, still has the problem of radioactive waste and weapons proliferation. It would be nice if this was not the case and scientists around the world have also researched a nuclear fusion reactor based upon helium-3. There are actually two different reactions that use helium-3. One is the deuterium/helium-3 and the other is helium-3/helium-3. In a deuterium/helium-3 reaction the production of neutrons is reduced by 80% and with a helium-3/helium-3 there are no extra neutrons created. This would eliminate radioactive contamination, eliminate the weapons proliferation concern, and if the reactor failed for some reason there would be no release of any radioactive gas (helium-3 is not radioactive and tritium is). It would seem that using these two fuels would make much more sense than using deuterium/tritium. However, there are several difficulties that make this not as easy as we would like it to be.

The nice thing about helium-3 is that one pound of it would give three to ten *million* times the energy as a ton of coal. However, (good news bad news again) deuterium-helium-3 reaction requires temperatures four times higher than the 100 million degrees required by the deuterium/tritium reaction. Even higher temperatures are required for a pure helium-3/helium-3 reaction. While there are solutions envisioned for these higher temperatures it is much easier to do the lower temperature fusion first then use the knowledge gained for designing reactors for the higher temperatures. Another problem is that there is very little helium-3 on the earth. This is where the Moon comes in.

Apollo 17 astronaut and geologist Harrison Schmitt estimates that even if helium-3 were to cost four billion dollars per ton it would cost the equivalent energy value for oil of about $28 dollars per barrel.[xii] Since, oil is up around $38 dollars a barrel today with no reduction in sight this could be a powerful incentive for lunar development. However, to get that one ton of helium-3 you would have to heat 20,000 tons of lunar soil (regolith) to a temperature of 700 degrees centigrade. It would take about 30 tons a year of helium-3 (assuming the reactors were designed, built, and up and running) to provide all of the energy needs of the United States. That is in contrast to the several *billion* barrels of oil that we use per year to generate electricity and there would be zero CO_2 pollution as well. It is important to remember here that the U.S. generates

20% of the total energy used on the planet so that 30 tons represents a large fraction of our planetary demand. Thirty tons of helium-3 per year at $4 billion dollars per ton would be $120 billion dollars per year in revenue for a lunar base.

The amount of helium-3 that would be required is not that much, many return systems could ferry that much helium-3 down per year quite easily or many times that. While processing 20,000 tons of regolith to get one ton of helium is a lot, that is quite well within what we do on the earth to mine precious metals. Fortunately, the Moon is not the only place to mine helium-3.

All of the four gas giants, Jupiter, Saturn, Uranus, Neptune, have incredibly large amounts of helium-3. The four gas giants have the same atmospheric composition as the original solar nebula where the sun and all of the planets formed. The Moon has helium-3 because it is a waste gas in the solar wind, left over from the sun's fusion reaction and the original primordial mix. The difficulty is that we don't know how to get a platform down into the atmosphere of Jupiter or Saturn and get it back out. Their escape velocities are 60, and 36 kilometers per second respectively. On the other hand the escape velocities of Uranus and Neptune are only about 20 kilometers per second, less than twice the escape velocity of the earth at 11.2 kilometers per second.[xiii]

A fission nuclear thermal rocket described earlier in this chapter has sufficient energy to get a "floater" helium-3 processing facility out of the atmosphere of Neptune or Uranus (you would use Uranus because it is a lot closer to the earth). There is no new physics that we have to learn, just implementing what we already know how to do. Dr. John Lewis of the University of Arizona has brilliantly thought this out in his book, *"Mining the Sky,"* a book that I will return to as we go forward. Dr. Lewis brings out that even if we used 4,000 tons a year of helium-3 from Uranus we would have a 4 *billion* year supply for our use and that is just in the upper part of the atmosphere. This amount is sixteen million times greater than the amount available on the Moon. It is ludicrous to believe that we are running out of energy and have to return to some state akin to the Middle Ages to survive.

As challenging as using helium-3 is, and as far as we may have to go to get it, there is a clear path to achieving the use of these materials to generate electrical energy. We understand the physics and we at least know what path to take to learn the engineering for helium-3 and we know today most of what we need to build a working deuterium-tritium fusion reactor. The only element that is lacking is funding. It is frustrating and infuriating to those of us who are space advocates who know that all of these things are possible but see billions spent in other areas and to be told that there is no money for space. In the end one of the reasons for this is connected to the mindset adopted by those who have faith in what was written in "The Limits to Growth." Nothing could be further from the truth and for the amount of money that we waste in most branches of government we could have the entire solar system at our command.

In truth it is the failure of the space advocacy community itself for not creating alliances with other special interests and postulating a space development effort that addresses the large problems of our day. Space advocates many times are their own worst enemies in that they see the vision of what is possible in space without considering the objections of other competing interests. It is often said that space advocates "preach to the choir," and this is very true. This will have to change if there is to be any true progress made and the solutions that are posited must be cost effective, require a minimum of government direction, and have as their goal a way to make a profit on the venture. Anything else will continually require begging for handouts from the government.

Energy Summary

In summary we have seen in this chapter, that from advanced solar arrays, to the power of the sun in helium-3 fusion, we have the potential not just to transcend the limits to growth but to blast them into oblivion. However, up until this point we have not addressed the companion issue of resources, that in the end, is intimately tied to energy as one of the things strongly addressed in "Limits to Growth," "Beyond the Limits," and "Earth in the Balance". The fundamental thesis of these books is geocentric, not accounting for the incredible bounty of energy and other resources of the solar system that can be used to enrich all of our lives on the earth. There is a saying in the environmental movement that goes, "Think Globally, Act Locally." This is no longer adequate if we are to survive and prosper into the 22nd century and beyond.

Looking beyond the earth I will address the hydrogen economy and how plentiful and portable energy is key to the freedom of movement that billions on the earth enjoy today and that we are desirous to extend to the rest of our brethren. Fusion and solar power is pretty much a fixed site power generation technology, not suitable for airplanes, cars, and other power hungry mobile applications. This is where the hydrogen economy comes in. The hydrogen economy is intimately intertwined with material resources as well as energy to make it happen so it will form the basis of our transition to the rest of the lunar development chain whose links we are still building.

*   *   *   *   *

[i] www.emcore.com, www.spectrolab.com
[ii] http://www.grc.nasa.gov/WWW/RT2001/5000/5410bailey1.html
[iii] Personal communication with Dr. Marzwell, 2-10-2004
[iv] http://www.thebulletin.org/research/qanda/reactors.html
[v] http://www.policyalmanac.org/environment/archive/nuclear_energy.shtml, Holt and Behrens, Congressional Research Service, July 2003
[vi] Ibid, http://www.policyalmanac.org/environment/archive/nuclear_energy.shtml
[vii] http://www.nei.org/index.asp?catnum=1&catid=5
[viii] *Hearings Before the Committee on Aeronautical and Space Sciences*, United States Senate, Eighty Ninth Congress, First Session on S.927
[ix] *Bring_the_Power_of_the_Stars_to_Earth.pdf*, U.S. Department of Energy, Office of Science Strategic Plan 2004,
[x] http://www.iter.org/
[xi] http://www-fusion-magnetique.cea.fr/gb/en_savoir_plus/lithium/lithium.htm
[xii] http://www.newhouse.com/archive/wylie012804.html
[xiii] *Mining The Sky*, John S. Lewis, Helix Books, Addison-Wesley, Reading MS, 1996

Chapter 6:

The Hydrogen Economy and Hope for the Future

In the previous chapter on energy, it was shown that, if we have the will to go forward with the technologies that we already know how to develop, there is no such thing as a shortage of energy. However, to make this energy available, we need to provide the funding necessary to bring these developments to fruition. Now, a hard-core free market proponent would say that this is not an area in which the government should be involved. Free enterprise will come up with what is necessary for us to transcend the limits to growth, but history shows that this is usually not the case.

It has been very proper throughout history for the government to invest the long-term capital necessary to bring entirely new economies into being and to support macroscale development. This goes as far back in history as Phoenician ships and Roman roads. The Phoenicians built an empire based upon trade and naval power that dominated the Mediterranean for hundreds of years and brought great wealth to their nation. This trading empire threatened the dominance of Rome—as the nation state of Carthage—and produced the great general Hannibal. The Romans expended tremendous resources on infrastructure. They built towns for trade and roads for the rapid deployment of their military power which helped the Roman Empire to last a thousand years.

In the modern age, this has been exemplified by state supported developments such as the "National Railroad," the Panama Canal, and the European and American Interstate Highway system. These modern investments have provided a tremendous return to the countries that have built them and every day, every one of us benefits from these investments; many built over a century ago. The investments by the Romans in roads helped to keep Europe together as a social if not political unit until the dawn of the modern era. Without such national investments in the development of fusion, the Hydrogen Economy, and the development of a space-based economy, we can rest assured that the scenario presented in, "The Limits to Growth," has a much higher probability of coming true in our lifetime.

It is unfortunate that, over time, most of the discussion about resources from space, and particularly the Moon, has not been based on those resources of immediate strategic value to the terrestrial economy. Geologists historically have been dismissive of such efforts, pointing to the fact that none of the mechanisms that concentrate ore on the earth work on the Moon. Indeed, many who advocate using resources from space ignore the Moon, or make comments such as, "the Moon is the slag heap of the solar system." This is based upon research on the materials brought back from the Moon by the Apollo program. These samples all indicate that, in the history of the Moon, there has never been any water. On the earth, water is a primary means for concentrating ores in an economical way, using existing extraction processes. However, this is not the whole story of lunar resources.

There are some resources on the Moon that are unique to that location such as helium-3 from the solar wind. The counter argument is that, if you had a billion tons of it and could ship it to the earth for free today, we could not use it in a fusion reactor because that reactor does not exist and will not exist for decades. This is a true statement and represents a large credibility problem for such mining business plans. The counter to this is that if we are going to have helium-3 fusion it behooves us to start the process of obtaining the fuel for it sometime soon. Just as a brief aside, the developments that I talk about here are not going to be built in a day or a year but themselves are the work of decades. The interstate highway system in the United States took almost 30 years from the signing of the Interstate and Defense Highway Act of 1956 until the system was completed in the early 80s.

This is not 40 years ago when the standard arguments against lunar ore formation were made. We know so much more about ore formation processes and how things like asteroid impacts on the earth have affected ore formation. Therefore, the goal should be to re-examine the Moon and see what could be used today, and how these resources could be used to address problems that exist on the earth. As an engineer or scientist it is always instructive once in a while to go back and question all of your assumptions. The failure to do this has been a tremendous impediment to progress throughout history.

If by some miracle, a resource could be found that also contributes to the solution of our energy and resource problem here on earth, then a compelling argument could be constructed for a "Return to the Moon," and the development of mining and other infrastructure there. The answer to this question is that there *are* resources on the Moon, the asteroids, and other locations, that meet this criterion, and these resources will be key factors in initiating the Hydrogen Economy.

The Hydrogen Economy

Fuel Cells vs. Internal Combustion Economics

There are actually two types of combustion engines that are prevalent in the world today. External combustion engines are steam engines, such as the one that Watt invented, that powered the industrial revolution and generate electrical power today. These engines burn a fuel, wood, coal, oil, or nuclear energy, which heats water external to the engine, creating steam to generate power. Internal combustion engines are typically gasoline, diesel, or turbine engines.

Since in previous chapters we have dealt with the issue of steam and internal combustion engines as well as nuclear energy for fixed power generation, I will concentrate on portable energy generation systems, such as internal combustion engines, that can't directly benefit from fusion or solar power. I will also ignore the turbine engine because it is limited to applications, such as airplanes, that can't benefit from fuel cells, at least in the near term.

In order to make a clear comparison between fuel cells and internal combustion engines, it is necessary to first explain how internal combustion engines and fuel cells work, and then compare the relative efficiencies and costs of each.

Internal Combustion

There are actually two types of internal combustion engines that deserve attention. The first is the Otto Cycle engine, originally developed by Nikolaus Otto in 1876. This is the familiar gasoline engine that is in most automobiles around the world. The second type is the familiar Diesel engine, invented by Rudolph Diesel in 1893. Diesel engines in the United States are predominantly used in large vehicles such as "18 wheelers," or large interstate cargo trucks. Diesels are also used in automobiles in the U.S., but make up a far lower percentage of vehicles on the road. Nearly 75% of new vehicles being sold today in Europe are Diesels. This is primarily because of the much higher cost of fuel in Europe relative to the U.S and the greater fuel economy of modern diesels.

Both of these engines are heat engines as they operate by the explosive combustion of gasoline or diesel fuel in a cylinder which then drives a piston that then uses the Watt patent idea of converting reciprocal motion to the rotary motion of a crankshaft. Figure 6.1 is a representation of an Otto Cycle single cylinder engine operational cycle:

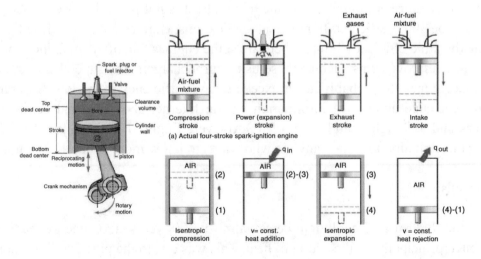

Figure 6.1: Gasoline Engine (Otto Cycle) Operation (Courtesy widget.ecn.purdue.edu)

In figure 6.1, the illustration begins with an air-fuel mixture that is compressed as the piston pushes upward. At the top of the piston stroke (Top Dead Center), a spark plug fires (Otto cycle), or the air-fuel mixture spontaneously ignites due to the pressure (Diesel cycle), pushing the piston down and turning the crankshaft. On the next stroke, the exhaust gases from the combustion are pushed out through the exhaust valve until the piston again reaches Top Dead Center. Then, as the piston moves back down, it pulls more air-fuel mixture into the cylinder via the intake valve. The cycle then repeats. This is how virtually all internal combustion engines work.

This is a hugely inefficient process that uses only a fraction of the total heat generated by the combustion cycle. In a gasoline engine, fully 50% of the heat is lost in the exhaust. Another 25% is lost in the cooling loop, and mechanical losses in the crankshaft, driveshaft, axle, and tires. This gives a normal efficiency of about 25%. Another way of looking at this is that, for a 200 horsepower engine, 400 horsepower is lost in the exhaust and another 200 horsepower is lost in the cooling and mechanical losses. A diesel is somewhat more efficient in its operation due to the higher compression ratio (a gas engine is usually 8:1-10:1, while a Diesel engine is 12:1-14:1) for the spontaneous ignition of the diesel fuel-air mixture. With these small design differences, a Diesel engine is about 37% efficient. The best stationary Diesel engines, running at a constant RPM and load, are about 54% efficient.[i] In comparison, an external combustion steam engine, such as those used for power plants, is about 30-35% efficient.

The second law of thermodynamics and the Carnot limit cannot be exceeded in a practical system. The Carnot limit, a formula that describes the maximum efficiency of an internal or external combustion engine, is the ratio of the temperature inside the engine to the temperature outside the engine. Since the outside temperature can never be absolute zero, and is usually equivalent to the outside air temperature, and the inside temperature is limited by engine materials, it is not possible for a combustion engine to be 100% efficient. In practical applications, engines today are close to their thermodynamic efficiency limits. There are some engines that offer small incremental efficiency improvements such as those using turbochargers and superchargers, but a gasoline engine will probably never exceed today's efficiency in the low 30% range. There is a huge amount of information on the Internet about internal combustion engines and a Google search term of, [Otto Cycle Engine Efficiency] will produce enough information to satiate anyone who wants to know more about these engines.

Fuel Cells

Since the Industrial Revolution began almost 300 years ago, energy has been generated principally by the action of heat. This was due to the practical limitation of direct hydropower (not enough rivers) for providing rotational energy to run industrial machinery. Power in the form of heat, was first generated by burning wood to boil water in order to provide steam power for machines. As the limits to growth for wood use rapidly came, coal became the fuel of choice for steam engines. Coal is an efficient and portable energy source that enabled the expansion of fixed steam engines for industry and the development of mobile steam engines that gave us paddle wheel steamers, steam locomotives, and even the Stanley Steamer automobile. When petroleum oil was discovered and brought forth in massive quantities, its superior heat capacity rapidly displaced coal in most areas, especially in transportation, where it sparked the automobile, air, and space age. The mighty Saturn V Moon rocket used nothing more than refined kerosene and oxygen (several tons per second) to lift seven million pounds of fuel, metal, and men from the surface of the Earth. All of these applications use combustion of fuel and oxygen to work their magic. However, combustion is inefficient, polluting, and life limiting to the machines that use it. There are other ways to generate energy that do not rely on combustion.

Over 150 years ago, an English lawyer turned scientist (a great rarity!), named William Grove, invented what was known then as the "Grove cell" or "gas battery" to generate electric current.[ii] This cell used a platinum electrode in a nitric acid bath, along with a zinc electrode in zinc sulfate, to generate 21.6 watts of power; a very respectable direct current power source for the time. Grove figured out that, by placing two platinum electrodes with one end of each in a container of sulfuric acid and the other ends separately sealed in containers of hydrogen and oxygen, then a current flowed between the electrodes. The sealed containers with the hydrogen and oxygen also had water in them and, as the current flowed, it was noted that the water level in the containers rose. This was the first fuel cell and it was based upon the chemical oxidization of hydrogen and oxygen. Other researchers investigated different types of fuel cells but none of them caught on due to the high cost and rarity of platinum.

In the late 1930s, Francis Thomas Bacon researched alkali electrolyte fuel cells and during World War II developed one that could be used in a Royal Navy submarine. He built other cells after the war, all the way through the 1950s, for the British National Research and Development Corporation.[iii] It was a descendant of these cells that supplied power for the Apollo Service Module that flew to the Moon a decade later. The Apollo program provided an influx of capital to develop fuel cells for the lunar program. They were already using liquid hydrogen and oxygen in the upper stages of the Saturn V and the "waste" product from fuel cells was ultra pure water for the crew to drink. Also, with the inefficient solar cells of the day and the weight of batteries, fuel cells became the simplest solution to a large number of problems besetting the Apollo engineers. On the other hand, dangers were evident when the explosion of one of the fuel cell oxygen tanks that caused the infamous "successful failure" of the Apollo 13 mission. Today in space, fuel cells are still used to provide power for the Space Shuttle for missions as long as 17 days. However, it is in more earth-bound applications where fuel cells are surging into mainstream use. Figure 6.2 shows how different types of modern fuel cells work:

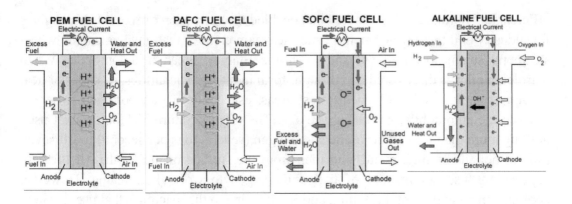

Figure 6.2: Various Hydrogen Fuel Cells Operation (Courtesy www.doe.gov)

In figure 6.2, the hydrogen molecules enter from the left and pass through an anode made of platinum catalyst where the electrons are stripped away. The positively charged hydrogen protons then pass through a chemical electrolyte containing water that does not allow the electrons to pass through, being an impermeable barrier to them. This is called a Polymer Electrolyte Membrane (PEM). The negatively charged electrons must flow around the membrane, creating an electrical circuit. On the cathode side of the fuel cell, oxygen molecules combine with the positive hydrogen ions and its wayward electrons, closing the circuit and making water and heat. The heat from a PEM fuel cell, such as the one above, is at about 80 degrees Centigrade or 176 degrees Fahrenheit. This is an electrochemical process that is not subject to the Carnot limit. There are several types of fuel cells available today and the type used depends on the application. Table 6.1 illustrates the types that are either in common operation or soon will be:

Type of Cell	Acronym	Operating Temp (C)
Direct Alcohol	DAFC	50-100
Polymer Electrolyte	PEM	50-100
Phosphoric Acid	PAFC	200
Alkaline	AFC	50-250
Molten Carbonate	MCFC	600
Solid Oxide	SOFC	500-1000

Table 6.1: Fuel Cell Types in Use Today

Each of these fuel cell types has their advantages and their market niche. The Solid Oxide and the Molten Carbonate types are best at generating stationary power. One of the nice things about them is that their waste heat can be used for secondary purposes such as running a steam turbine or to provide heat for a building. This raises the overall efficiency higher than for the lower temperature fuel cells. The Polymer Electrolyte seems to be the best candidate for medium power applications such as automobiles. Right now, intensive research and pre-commercial development are under way for the Direct Alcohol cells that can be used to power cell phones, PDA's and Laptops. These are the leading contenders today for commercial applications, but research is ongoing and the lead is subject to change. All of these cells are very efficient in comparison to batteries, or to combustion processes, which is the primary factor driving interest in them today.

The measure of a fuel cell's thermodynamic efficiency is given by the ratio of the Gibbs function change to the Enthalpy change in cell reaction. The Gibbs function change measures the electrical work and the enthalpy change is a measure of the heating value of the fuel. This of course is Greek to those who are not scientists but suffice to say that fuel cells are always more efficient than internal combustion processes. Without getting into the mathematics, the ratio above translates into a maximum possible efficiency of 83%.[iv] In practice, this is not achieved, but efficiencies far higher than that possible in combustion engines are realizable. This is the technical advantage of fuel cells that gives them their advantage over combustion engines.

Fuel Cells vs. Combustion Efficiency

In comparison to combustion processes, fuel cells deliver a considerable improvement in the conversion of energy into useful work. This is across all applications, from stationary power plants, to portable applications for transportation. This advantage can be further extended by the reuse of the heat generated by fuel cells to turn traditional power generation turbines, increasing the total system efficiency. A downside to fuel cells in operation is that they require hydrogen to work. Since hydrogen does not exist independently from other atoms in nature, it has to be produced by some method. This method is called hydrogen reformation and requires an input of energy to produce hydrogen. The methods required to "re-form" hydrogen bearing molecules to obtain pure hydrogen diminish some of the advantages of fuel cells, but even with these disadvantages, fuel cells are a superior technical solution over combustion engines. Table 6.2 gives some comparisons between fuel cells and combustion engines:

Power Source	% Ideal Efficiency	% Efficiency Practical	With Reformation	Comment
Steam (External)				Power Plants
Otto Cycle	59[v]	25-32[vi]	N/A	Gas Engine
Diesel Cycle	56[vii]	52[viii]	N/A	35% Typical
Carnot (Ideal)	85[ix]	N/A	N/A	Not Realizable
Atkinson Cycle	59	34[x]	N/A	Hybrid Engine
Gas Turbine	59	57	N/A	Power Plants
PEM Fuel Cell	85[xi]	60	40[xii]	Water Electrolyte
High Temp Fuel Cell	70	55	52[xiii]	Carbonate or other Electrolyte

Table 6.2: Combustion Vs Fuel Cell Efficiency Comparison

For completeness and to further illustrate the advantages of fuel cells, I added the Atkinson Cycle internal combustion engine that is used in hybrid vehicles like the Toyota Prius, the gas turbine and a high temp fuel cell used for stationary applications. Figure 6.3 is a graph of relative efficiency of a 2.5 megawatt Molten Carbonate ship based fuel cell versus a diesel or gas turbine generator:

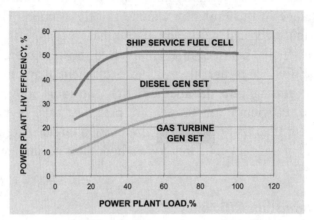

Figure 6.3: Efficiency vs. Load Comparison for Fuel Cell, Diesel and Gas Turbines

(Chart Courtesy Fuel Cell Energy Inc.)

In figure 6.3 above, the efficiency values are for actual end-to-end power generation, using the same fuel, the same loads and the same operating conditions. This provides a realistic scenario for comparison purposes. This also eliminates any discussion about the merits of hydrogen production to feed the fuel cells because this is also incorporated in the efficiency calculation. The operating temperature for this particular fuel cell is 650 degrees centigrade (1202 F) and the waste heat is used for the steam reformer for the fuel and for general heating uses on board the ship. This system is in commercial production by FuelCell Energy Inc, with support from the U.S. Navy.[xiv]

In practice today, fuel cell efficiencies are highly variable, depending upon the fuel (hydrogen or hydrocarbons with re-formation), and the type of catalyst used. The use of waste heat in a fuel cell also effects efficiency in a positive way. Often that waste heat is used to turn a steam turbine, or to heat the premises where the fuel cell is deployed. There is also variability in the efficiency of internal combustion engines, depending upon whether or not they are used in stationary applications with constant power (diesel generators), or in automobiles (gas or diesel). In automobiles maximum efficiency is rarely reached. This is why fuel economy is so much worse in city driving compared to the highway.

The Hydrogen Economy and Pollution

While fuel cells are considerably more efficient at turning energy into power than their combustion counterparts, their real potential for a revolution is in their low output of pollutants. In a fuel cell that uses hydrogen, the only output is pure water. This water is more pure than the best water from a municipal water system. Water output from fuel cells was used for drinking water by the crew of the Apollo missions to the Moon as well as on the U.S. Space Shuttle. Figure 6.4 shows the relative outputs of pollutants from the FuelCell Energy Inc. 2.5 megawatt system described above.

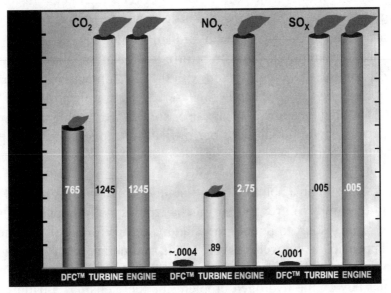

Figure 6.4: Global Warming Gas Outputs from
Diesel and Gas Turbines Vs Fuel Cells

In the scientific community, one of the measures of anthropogenic (man made) climate change relates to relative concentrations of CO_2 in the atmosphere. While the debate about whether or not CO_2 is a first order contributor to the current global warming trend continues to rage, fuel cells do offer an option to dramatically reduce the amount of CO_2 pumped into the atmosphere without having to give up our mobile society, whether or not global warming from CO_2 is truly a problem. Also, with the limited remaining supply of oil, hydrogen fuel cells are the only long-term solution to provide energy for transportation in a much more efficient manner than internal combustion engines. Another interesting aspect is the wild card of methane ice. Methane is much easier to reform and produces far less CO_2 than other heavier hydrocarbons. Therefore a compelling argument can be made that, even if we had all the oil possible to want, that we would want to use fuel cells and light hydrocarbons like methane.

In figure 6.4 above, the CO_2 output of the fuel cell is less than 60% of the diesel and gas turbine generators, even using exactly the same fuel. The same savings are typical with all types fuel cells, used in both fixed and mobile applications. For fuel cells that use hydrogen alone, the CO_2 output is zero. However, there are only two processes to produce hydrogen: one is the steam reforming of hydrocarbons that uses energy and emits CO_2, the second is direct electrolysis from solar power and nuclear energy that emits zero CO_2. This is the quandary of the environmental movement today: go to the hydrogen economy and only cut the proportion of CO_2 by half, or support nuclear power (fusion preferably) to reach zero emissions.

This is where we begin to see the convergence between Fuel cells, fusion and solar energy. Using fusion, you get zero CO_2 and little or no radioactive output. The power generated can be used to directly make hydrogen from water, producing oxygen as well. The heat can also be used to steam reform hydrocarbons into hydrogen. Direct solar energy can be used if we go forward with the development of the Quantum Dot cell. Direct solar can be used, especially in desert areas, for electrolysis of water into hydrogen for fuel or for power. This also solves the problem with hydrogen distribution, because hydrogen is far more difficult to move through pipelines than methane or Liquid Natural Gas.

In the end, it is going to be a combination of solutions that take us beyond the end of the oil age and the limits to growth that this has traditionally implied. However, the limits to growth for resources remain to be addressed. With all of their advantages, why have fuel cells not taken over the world? It is a combination of factors, but, in the end, it all comes down to cost and the intrinsic rarity of platinum on the Earth. With fusion and solar power, the hydrogen problem for fuel cells has a pathway to being solved, but there is another problem. That problem is the platinum catalyst that almost all fuel cells use. Without platinum and other related metals, the Hydrogen Economy cannot come into being because there are no known catalysts with similar properties. Therefore, our attention turns to this resource: a most important one for our future.

*   *   *   *   *

[i] http://hetexengines.com/diesels.shtml

[ii] http://fuelcells.si.edu/origins/origins.htm

[iii] Ibid

[iv] http://www.princeton.edu/~humcomp/sophlab/ther_58.htm

[v] http://ecen.com/content/eee7/motoref.htm

[vi] Ibid

[vii] http://hyperphysics.phy-astr.gsu.edu/hbase/thermo/diesel.html#c1

[viii] http://ecen.com/content/eee7/motoref.htm

[ix] http://hyperphysics.phy-astr.gsu.edu/hbase/thermo/carnot.html

[x] http://home.earthlink.net/~graham1/MyToyotaPrius/Understanding/InternalCombustion.htm

[xi] http://www.eere.energy.gov/hydrogenandfuelcells/fuelcells/why.html

[xii] Ibid

[xiii] *Development of a Ship Service Fuel Cell*, Ghezel-Ayagh, et al, AES 2000/All Electric Ships, October 26-27, 2000, Paris France

[xiv] Ibid

Chapter 7:

The Hydrogen Economy and Platinum

Platinum, The Key to, and the Achilles Heel of, the Hydrogen Economy

Platinum and other Platinum Group Metals (PGM's: platinum, iridium, osmium, rhodium, ruthenium and palladium) are both the key to, and the Achilles heel of, the Hydrogen Economy. PGMs are extremely rare in the Earth's crust, making up only a few parts per billion in igneous rocks. There are certain areas of the world where PGMs are found in concentrations that make mining profitable, but it has to be pointed out that these concentrations are still very small. In the best mines in the world, located at the Merensky Reef in the Bushveld Complex in South Africa, the concentration of PGMs is only 7 to 9 grams per ton with a global average of 4 gm/ton. Today, a fuel cell that runs a small car generates about 50 kilowatts of power. This power level takes about 2 ounces, or 57 grams of platinum for the catalyst. That means that, in order to obtain enough platinum for one car, miners have to dig fourteen tons of ore. Multiply this by the estimated 3 billion cars that a moderately affluent world would have in 2050 and the picture becomes clear.

Research is underway to lower the platinum loading (the amount of platinum per fuel cell) to only about 0.2-0.3 ounces or ~5.7-8.5 grams of platinum per automobile.[i] It is hoped that this level of platinum loading will be reached by 2020 for small cars. An SUV class vehicle or freight hauling tractor-trailer would take many times that amount of platinum. There is some question about the ability to reach these low quantities of platinum for an automobile fuel cell stack but the trend is positive. Recently, laboratory researchers have achieved 22.6 gram, (0.8 ounce) platinum loading. This is a subject of intense research since this is a crucial factor in lowering the cost of a fuel cell powered automobile. Table 7.1 gives the world mine production, reserves and reserve base of platinum and palladium; the two most used PGMs:

| Location | Mine Production (Kilograms) | | | | PGM (Kilograms) | |
| | Platinum | | Palladium | | Reserves | Base |
	2002	2003	2002	2003		
USA	4,390	4,100	14,800	14,600	900,000	2,000,000
Canada	7,000	7,000	11,500	11,000	310,000	390,000
Russia	35,000	36,000	84,000	74,000	6,200,000	6,600,000
S Africa	134,000	135,000	64,000	64,800	63,000,000	70,000,000
Other	3,400	5,000	6,900	7,000	800,000	850,000
Total	*184,000*	*187,000*	*181,000*	*171,000*	*71,000,000*	*80,000,000*

Table 7.1: World Production/Reserves and Reserve Base for Platinum and Palladium[ii]

According to the US Geological Survey, South Africa has the largest reserves of platinum as shown in table 7.1. In declining order, this is followed by Russia, the U.S.

and Canada. Other minor producers are lumped in as "other." The USGS estimates that the total global reserves that can be mined on the Earth are about 100 million kilograms. If 3 billion cars (an estimate based upon current trends) will be on the road by 2050, the total global reserves of platinum would be sufficient to power these cars if the platinum loading of 5.7 grams (0.2 ounces) is reached. If this level is not reached or if you include other uses of platinum and other PGMs, it seems that planetary reserves may not be sufficient to support the full transition to the Hydrogen Economy. In addition to the questions of reserves, there is the issue of the environmental costs associated with the extraction of minute amounts of platinum and other PGMs from tons of ore. With the amount of PGMs necessary to fuel the Hydrogen Economy this is not a trivial issue.

Here is another situation where there is a divergence of view concerning what the reserves of platinum are just like what we saw with oil. In a 2002 presentation by Dr. Gerhard von Grunwaldt, Vice President of the South African National Research Foundation, he presented numbers on the PGM reserves that are substantially different to the USGS numbers. Figure 7.1 gives the number for proven and unproven reserves of platinum:

Reef Type	Proven and Probable Reserves	Inferred Resources	Total Resources
Merensky Reef:			
Eastern Bushveld	11	286	297
Western Bushveld	66	114	180
UG-2:			
Eastern Bushveld	38	306	344
Western Bushveld	78	96	174
Platreef	10	136	146
TOTAL	203	938	1141

Figure 7.1: South African Government Estimates of Platinum Reserves (ounces)

In Table 7.1 the USGS numbers for reserves for South Africa at 63 million kilograms and the Reserve Base at 70 million kilograms. Figure 7.1 from the South African government contradicts this as they estimate reserves at 203 million ounces or 5.7 million kilograms and a reserve base (inferred resources) at 938 million ounces or 26.5 million kilograms. The total South African economically viable resources are 1141 million ounces or 32.2 million kilograms. This is less than 10% of the resources estimated by the USGS for proven reserves and together less than half of the reserve base. These numbers are considerably smaller than those of the USGS and are cause for considerable concern.

The South Africans have a similar view of global platinum reserves. Figure 7.2 gives the South African estimate of global platinum reserves.

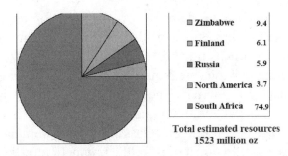

☐ Zimbabwe	9.4
☐ Finland	6.1
■ Russia	5.9
☐ North America	3.7
■ South Africa	74.9

Total estimated resources
1523 million oz

Figure 7.2: South African Estimate of Global Platinum Reserves

The South African estimate in figure 7.2 above of global resources is 1523 million ounces or 43.3 million kilograms. Again this is about half the estimate of the USGS. Who is right in these estimates? This is a key question that will ultimately govern the cost of platinum and our ability to make the switch to the Hydrogen Economy if we are to rely solely on terrestrial sources of platinum. Indeed this may drive us inevitably to extraterrestrial resources if the South African estimates are correct.

Platinum Usage Trends in the Transition to the Hydrogen Economy

The UK government Department of Transportation commissioned a brilliant study, executed by AEA Technologies, entitled, *"Platinum and Hydrogen for Fuel Cell Vehicles."* It is an extensive treatment of the issues surrounding the global transition to the Hydrogen Economy.[iii] This study covers all of the relevant parameters concerning the production and use of platinum today for transportation, along with future demand as we move toward the hydrogen economy.

According to the study, even without the hydrogen economy, the transportation industry uses a lot of platinum and palladium. As of 2002, the automotive industry used about 70,750 kilograms of platinum and palladium, equal to 20% of global production.[iv] This is expected to increase with more stringent pollution controls on diesel automobile engines in Europe and North America. Platinum is a valuable commodity in applications beyond fuel cells and catalytic converters. Table 7.2 illustrates some of the principal uses of platinum (thousands of ounces):

Sector	1994	1995	1996	1997	1998	1999	2000	2001
Automobile	1870	1850	1880	1830	1800	1610	1890	2520
Chemical	190	215	230	235	280	320	295	290
Electrical	185	240	275	305	300	370	455	385
Glass	160	225	255	265	220	200	255	285
Jewelry	1740	1810	1990	2160	2430	2880	2830	2550
Petroleum	90	120	185	125	125	115	110	125
Other	585	570	495	535	620	515	515	515
Recycled	(290)	(320)	(350)	(370)	(405)	(420)	(470)	(520)
Total	**4530**	**4710**	**4960**	**5130**	**5370**	**5590**	**5880**	**6150**

Table 7.2: Global Platinum Demand by Sector (1994-2001) (Johnson Matthey 2002)

Especially in Asia, there is huge demand for platinum for jewelry because of its beauty and durability. Platinum is one of the most valuable metals, not just for its rarity and beauty, but also for its practical applications. The petroleum industry uses platinum in the catalytic cracking (breaking down of heavy hydrocarbons into lighter ones) of hydrocarbons in refineries. The electronics industry is using increasing amounts of platinum and palladium in the manufacture of hard disk drives and capacitors. In the electronics related glass industry, demand for platinum is accelerating since it is a required material for the production of these wonderful LCD screens that we have come to know and covet. The chemical industry uses platinum as a catalyst to lower the energy required for a plethora of chemical reactions, especially the production of silicone. In the "other" category above are things like platinum fillings, spark plugs, pacemakers, catheters, and many other items that need a high temperature or a corrosion resistant metal. Demand will soar for this versatile metal as we move toward the Hydrogen Economy.

In the future, the fuel cell industry's demand for platinum and other PGMs will dwarf all other sectors and will place an incredible strain on the supply of platinum and the environment. The production of platinum is a toxic enterprise, using tremendous amounts of chlorine, ammonia, and hydrogen chloride gas, which are all released as part of the process. Incredible amounts of effluents are left at the end of the process; several hundred pounds of effluent per gram of platinum and other PGMs. This toxic effluent contains metals such as iron, zinc, nickel, as well as other metals that were part of the ore, but not commercially viable to extract. This process also generates sulfates as a waste product. While major producers are using more environmentally conscious means to mitigate pollution, the sheer volume of material involved and the minute quantity of valuable PGMs per ton of ore, makes a clean process impossible.

A German study (Okoinstitute 1997), referenced in the UK government study, indicated that it takes 23.76 kilowatt hours of electricity per gram of platinum produced and 10.45 megajoules of natural gas. Based upon electricity generation and natural gas used, this translates into about 6.4 kilograms of CO_2 per gram of platinum.[v] This will vary widely based upon the source of electricity. In Ontario at Sudbury, the source may be hydropower, and in Russia, it may be from nuclear power, so accurate measures of total CO_2 per gram of produced platinum on a global basis is almost impossible to determine. However, based upon the above numbers for production in South Africa, we can estimate the total natural gas and electricity needed for the dramatic increase in production forecast in order to support the transition to the Hydrogen Economy.

Platinum Future Production Estimates

The US Geological Survey estimate of platinum production in 2003, listed in table 7.1, was 187,000 kilograms or about 6.6 million ounces. Demand has accelerated since the late nineties after a steady increase over a 20 year period from 2.5 million ounces a year in 1984. The UK study developed three different scenarios for the platinum demand curve based on different fuel cell adoption rates for transportation. Table 7.3 illustrates the adoption rates implicit in the three scenarios in percentage penetration as a function of new car sales:[vi]

Scenarios	2010	2020	2030	2040	2050
Slow	0%	2%	5%	15%	20%
Realistic	0%	5%	30%	70%	80%
Rapid	10%	50%	100%	100%	100%

Table 7.3: Fuel Cell Penetration Scenarios for New Car Sales (From UK Report)

The Study took these fuel cell penetration rate scenarios and then derived a platinum demand curve from that. Before we look at the demand curve, it is instructive to look at the U.S. government's projections in this area from the U.S. Department of Energy's imposingly titled, *"Platinum Posture Plan, an Integrated Research, Development, and Demonstration Plan,"* published in February 2004. This is part of the Bush administrations $1.6 billion dollar Hydrogen Economy implementation effort. The schedule in this document for fuel cell adoption is at the high end of the UK government's study. Figure 7.3 is from the DOE document:[vii]

Government-Industry Roles in the Transition to a Hydrogen Economy

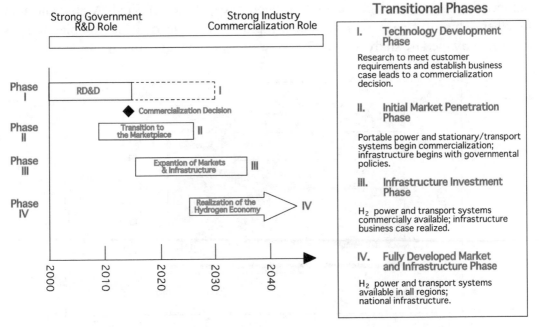

Figure 7.3: U.S. DOE Schedule for Hydrogen Economy Implementation

This schedule implies a "full steam ahead" approach that will lead to a full-scale adoption of the hydrogen economy in the 2030-2040 timeframe. This is illustrated in figure 7.3, which shows the DOE estimate of the decrease in oil consumption for transportation as we make the switch to hydrogen.[viii]

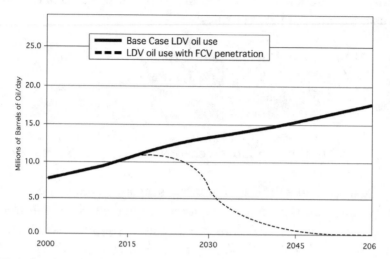

Figure 7.4: Projected Oil Demand Curve for Transportation With Fuel Cells

In an amazing statement, the DOE projects the *full* implementation of the Hydrogen Economy, with 100% penetration in the light duty vehicle segment, by 2038! It is very interesting to look at the convergence in ideas between the UK Department of Transportation's estimates and the U.S. DOE numbers. Interestingly, the DOE is estimating a savings of up to 11 million barrels of oil *per day* by 2047; an incredible number that is greater than the total production capability of Saudi Arabia and Iraq together. This of course will lead to a rapid increase in demand for platinum as estimated by the UK study. Figure 7.5 shows this estimated demand curve:

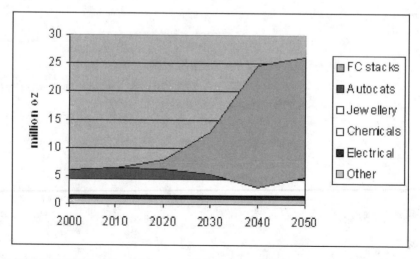

Figure 7.5: High End Platinum Demand Curve for The Hydrogen Economy

While the demand curve above has some problems (it assumes an unexplained dip in the demand for jewelry), it seems to be accurate overall in the total increase in demand with one really big exception. The UK study here is assuming a platinum loading for an automobile at 5.7 grams (0.2 ounces). This is a dangerous assumption since, today, it is at 2 ounces for a 50-kilowatt fuel cell stack. If this is off by even a factor of 2 (11 grams or 0.4 ounces), the demand curve could exceed 50 million ounces (1,415,000 kilos!) per year. Even if you split the difference, we are still talking about a million kilos per year just for this application. The total estimated planetary reserves are 100 million kilos (or 43 million if the South Africans are right), but what would be the environmental cost of this mining?

Using a conservative global average of 4 grams per ton, this means that, with a total demand of about 1.5 million kilos per year, we are looking at the following numbers.

CO_2 Per Year	6.4 kg/gm X 1.5 billion gm = 9.6 Million Tons
Kw/hr Electricity	23.76 Kw/hr X 1.5 billion gm = 35,640 Gw/hr
Natural Gas	10.45 Mjoule X 1.5 billion gm = 15,675 Tera-joules
Waste Material	4 gm/ton X 1.5 billion gm = 375 million tons

These are staggering numbers, except for the CO_2 production, which is a net gain over using oil, but is still 0.2% of the total 2004 CO_2 global emissions of 5.8 billion tons. The figure for natural gas is about 1% of the total natural gas consumption of the U.S. The electricity number of 23.76 Kw/hr translates into a power demand of about ~4 gigawatts per year. This is the equivalent of two large nuclear plants or enough energy to power about 2.5 million homes. This says nothing about the toxic waste that comes from the mining and processing of platinum and the total environmental effects in countries like South Africa and Russia which lack the stringent level of environmental laws such as are the norm in the U.S. and Canada.

Platinum and the Moon

It is my contention that we can get platinum and other PGMs from the Moon in quantities well in excess of the known reserves of the Earth. These resources are derived from the impacts of metal asteroids on the Moon's surface. At this time, it makes more sense to obtain the PGMs from the Moon rather than the Asteroids due to the long round trip times and the technological difficulty of operating that far from the Earth for extended periods of time with the technology that we have today. After a robust lunar infrastructure is operating, it will then become feasible and profitable to go after the asteroidal resources, but not before.

If we can shift the production of these high value metals off planet, we can have a material effect on the quality of life and the environment in South Africa, Russia, and any other location where PGMs are mined. This can be the starting point for developing the resources of the solar system for the benefit of the earth and it becomes a powerful argument for this development. I have coined a phrase for this development that works in the context of the new exploration vision laid out by

President Bush.

We go to Mars to extend our civilization there,
We go to the Moon to save our civilization here.

The exploration vision can be expanded to encompass the development of the resources of the Moon, especially the platinum and other PGMs necessary to enable our civilization to permanently transcend the limits to growth as outlined earlier. What evidence do we have that these materials exist on the Moon and in economically viable concentrations and quantities? The answer is that we have a lot of evidence and it comes from our study of meteorites, spectrographic studies of Near Earth Asteroids (NEAs), and the study of impacts and resources derived from impact sites on the Earth.

The Riches of the Solar System Await

In 1996, Dr. John Lewis, a planetologist from the University of Arizona, wrote an incredible book called, "Mining the Sky." In this book, Dr. Lewis lays bare the fallacy that we live in a resource constrained world and solar system. The difficulty is that, only now do we have the technological wherewithal to actually develop these resources. Also, it is only within the last 40 years that we have known of the promise of vast resources, as well as the enormous danger from Near Earth Asteroids (NEAs). In 1963, we only knew of 14 NEAs. Today, we have located over a thousand that are larger than half a mile in diameter, that have the potential to destroy our species just as effectively as the dinosaurs were 65 million years ago. At this point, a paragraph from Dr. Lewis's, "Mining the Sky," is appropriate:

> *...**Shortage of resources is not a fact; it is an illusion born of ignorance**. Scientifically and technically feasible improvements in launch vehicles will make departure from Earth easy and inexpensive. Once we have a foothold in space, the mass of the asteroid belt will be at our disposal, permitting us to provide for the material needs of a million times as many people as Earth can hold... Using less than 1 percent of the helium-3 energy resources of Uranus and Neptune for fusion propulsion, we could send a billion interstellar arks, each containing a billion people, to the stars. There are about a billion Sun-like stars in our galaxy. **We have the resources to colonize the entire Milky Way....***

I like it when a guy has ambition! Dr. Lewis's, "Mining the Sky," illustrates the huge gulf that separates the positive worldview held by space advocates and the limited worldview espoused in, "The Limits to Growth," "Beyond The Limits," and, "Earth in the Balance." I want to reiterate here that I do not believe that the people who wrote, or those who believe in the point of view espoused in the "Limits to Growth" and its successors are evil. They have looked at the data that they understand and have applied computer modeling to arrive at their results. They have developed from that study, a course of action that they believe is in the best interests of the people of the planet. It is as much the fault of space advocates as anything else since they have not put forth the arguments in a way that will carry the day. Who is right? The lives and happiness of billions of people and the future of human civilization depend on the answer to this question. The answer is both eye opening and hopeful for our future.

*   *   *   *   *

[i] http://www.dft.gov.uk/stellent/groups/dft_control/documents/homepage/dft_home_page.hcsp

[ii] http://minerals.usgs.gov/minerals/pubs/commodity/platinum/, Platinum Yearbook, 2003, (pgmmcs04.pdf)

[iii] *Platinum and Hydrogen for Fuel Cell Vehicles,* AEA Technologies Report for the UK Department of Transportation, 2002 (http://www.dft.gov.uk)

[iv] Ibid, P. 7

[v] Ibid, P. 12

[vi] Ibid, P. 16

[vii] *Hydrogen Posture Plan, An Integrated Research, Development, and Demonstration Plan*, United States Department of Energy, February 2004 P. 6

[viii] Ibid, P. 37

Chapter 8:

We Have Always Used Extraterrestrial Metals

Mankind has known of and used meteoritic iron for thousands of years. In looking at the history of metal in ancient civilizations, in almost every case, evidence is found of the use of iron from meteorites. Meteoritic Iron jewelry and tools older than 6000 years have been found in Iran, in the ancient cities of Uruk and Ur, as well as younger finds at Alaca Huyuk in Anatolia, and at Gerzeh and Dier el Bahari in Egypt.[i] In China, meteoritic iron implements have been dated to the beginning of the Chou period, around 1030 B.C. In almost every civilization around the world, evidence has been found of the use of meteoritic iron.[ii] In 1892, Admiral Peary found that the Inuit's used iron spear tips for hunting. When he inquired as to the source of the metal, he was led to a large meteorite at Cape York in Greenland. Today, this thirty-four ton hunk of Nickel/Iron meteorite resides at the Hayden Planetarium in New York. An even larger seventy ton piece is still in the ground at Groontfontein South Africa.

From the study of these meteorites, we know a lot about their composition and, surprisingly, the metal content varies quite a bit. The dominant element in these metal (M class) meteorites is iron. The second most common element is nickel, comprising 7% to 20% by weight.[iii] Cobalt, usually comprising about 1%, is the remaining major constituent of these M class bodies. They also contain a few tens of grams of PGMs for each ton of mass with a direct correlation to nickel concentration. A subgroup, called nickel rich Ataxites, can have up to 60% nickel. These are very rare however.

It is extremely interesting to note that, both in South Africa and in Canada, the PGMs mined are derived from the postulated impact of an asteroid on the Earth. Copper was discovered in the Sudbury area, (now known to be an asteroid impact site) near Ontario, Canada, in 1883. Since that time, over 20 billion pounds of copper and 20 billion pounds of nickel have been mined in the core of the Sudbury impact site.[iv] Along with the copper and nickel, over 20 million ounces of platinum and palladium have been mined. Figure 8.1 shows a map of the Sudbury impact and where mining occurs around the impact site:

Although it was hotly debated and even laughed at when evidence was first presented by the planetologist, Robert Dietz, in 1963, it is now recognized that the richness in metals found at Sudbury came from what Dietz called an "Astrobleme" ("star wound" in French).[v] The difference between an Astrobleme and a crater is that the former is a mere remnant of the original crater after most of it has eroded away. The impactor that slammed into the Canadian wilderness 1.87 billion years ago is estimated to have been 9-10 kilometers in diameter. This is comparable in size to the rock that killed the dinosaurs. However, that far back in time, all that would have been killed are a lot of microbes. The distribution of mines around the impact remnant at Sudbury is also remarkable. They are not in the center as one might expect, but around the edges of the site.

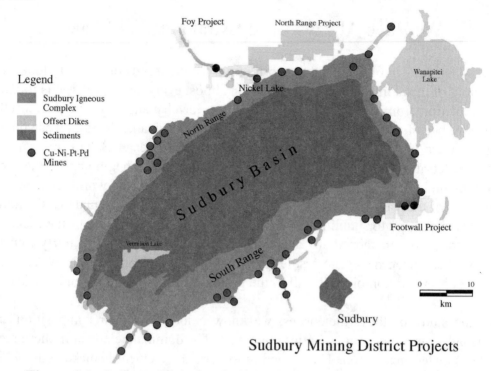

Sudbury Mining District Projects

Figure 8.1: Sudbury Nickel and Platinum Group Metal Mining District

In South Africa, it seems that the Merensky Reef in the Bushveld Complex is part of the Vredefort impact crater, estimated to be 2 billion years old and 300 kilometers in diameter.[vi] Figure 8.2 is a picture taken from the Space Shuttle of the Vredefort impact site and Astrobleme:

Figure 8.2: Vredefort Impact Site, South Africa (Picture Courtesy NASA).

The richness of the ores and their location plausibly tie them to the asteroid impact body. The link referenced for the Vredefort impact has some very good graphics and other information for follow up reading on this subject. The Earth Impact Database referenced here also has an extensive bibliography on this subject. One very good paper tying the Vredefort astrobleme to the Bushveld complex was written by Dr. Wolfgang Elston. In it, he discusses the possibility that the volcanism that produced the Bushveld complex was a result of the Vredefort impact.[vii] The Bushveld complex and its associated substructure, the Merensky reef, contains the vast majority of Southern Africa's PGM resources. Since the existence of Nickel/Iron and PGM enriched asteroids are well established, there is considerable evidence to support the thesis that the richest nickel (Sudbury) and PGM (Vredefort/Bushveld) mines in the world are tied to impact bodies.

This thesis has been bolstered by other recent research. In an article by the British Broadcasting Company (BBC) last November this connection was brought out by geologist Dr. Adrian Jones from University College, London:

Dr Jones added that one recently discovered major nickel deposit in Russia - coupled with two other, previous finds - suggested that some metals might come from the impactors themselves.

"It makes it rather interesting that two or three large impact structures are now associated with the same association of nickel-rich metals," he stated.

"The idea from our modeling and our smaller experiments [is] that the impact crater itself may still retain a mixture of materials, both from the melted crust and from the residue of the meteorite impact that has been redistributed around the crater.[viii]

"That would contain a lot of nickel-rich metals and platinum-group elements."

The Canadian government estimates that the total value of all of the metal bearing minerals (nickel, copper, PGMs) at the Sudbury impact site have a total estimated value of approximately $300 ***billion*** dollars.

Some of these impacts and resources can be from types of asteroids other than M class. Enstatite Chondrites have up to 25% nickel/iron and some of them have a metallic phase with little iron and as high as 80% nickel and PGMs.[ix] Other metals, such as titanium, chromium, copper, can be found in various quantities in various meteorite classes. If you take the thesis that meteorites are the materials left over from the formation of the solar system, the Earth itself is one great, big, huge asteroid that happens to be large enough to hold an atmosphere, water and life.

At the present time, we know the whereabouts of almost 3000 Near Earth Asteroids out of a total population estimated in the millions. Most of those that we know about are the larger ones that are a kilometer or more in diameter and have the capacity to wipe out our civilization. The amount of smaller ones goes up exponentially all the way down to dust grains that we see in the night sky as shooting stars. This is just the NEAs, not the greater population from the main belt in the region between Mars and Jupiter. Figure 8.3 are pictures of some of the known asteroids from the Galileo and NEAR spacecraft:

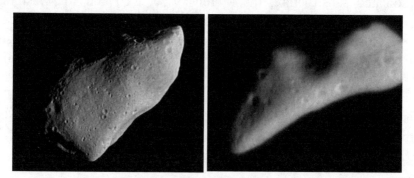

Figure 8.3: Gaspra and Eros, Two Near Earth Asteroids

The Minor Planet Center, (http://cfa-www.harvard.edu/iau/lists/Unusual.html), run by Dr. Brian Marsden, is the global clearing house for information on the orbits of these bodies. Beyond the NEA resources, in the main asteroid belt there are hundreds of millions of asteroids with only a fraction of them cataloged. In the outer part of the solar system, the number of these bodies is also very large. It has only been recently that significant numbers of these have been found. These however, have great resource potential in the far future for the vast amounts of water that we know from studying their spectra. From our study of meteorites and from observing meteors entering the atmosphere, we know that approximately 3% of all of the NEA population are nickel/iron bodies. Another 1% are a mixture of iron and rock, called stony irons.[x] These are the only ones of interest for discoverable resources since they have enough material strength to not completely vaporize on impact with the lunar surface. What this means is that, out of the roughly 3000 NEAs known today, 90 of them are pure nickel/iron with another 30 being stony irons. This is a fraction of the total that are out there.

One NEA that we know about is called 3554 Amun. This is the *smallest* known M class asteroid. It is approximately 2 kilometers in diameter. It also contains more iron, nickel, cobalt and PGMs than have been mined in the history of mankind. Dr. Lewis estimates that the iron and nickel in 3554 Amun is worth about $8 *trillion* dollars. The cobalt is worth another $6 *trillion* dollars. The PGMs and gold contained in this object adds another $6 *trillion* dollars to the total. This is a grand total of $20 *trillion* dollars worth of metal, $6 *trillion* dollars of that being enough PGMs to completely enable the switch to the Hydrogen Economy with enough left over to adorn every person on earth with a nice PGM ring![xi] While it is unrealistic to expect that a flood of metals on the market would keep the market value this high, the purpose of illustrating these numbers is that the amount of metals obtainable from Amun exceeds the total available Earthly supply.

Now, lest you think that I am advocating going to Amun or another asteroid first, look up in the sky on a clear night at the Moon. Four out of every 100 craters on the Moon were made by asteroids of the same material as Amun! If we used the nickel and cobalt from Amun, we could make every single bridge in the United States out of a type of stainless steel that would give them the natural corrosion protection to last thousands of years. If the Golden Gate Bridge were made of this material it would never need to be painted, saving a tremendous amount of maintenance costs. This would require importing tens of thousands of tons of nickel and cobalt to the Earth, a non trivial task today, but this is a desirable action as we march into the future.

Amun is the smallest identified nickel/iron NEA. The largest that we know of is the main asteroid belt object called 216 Cleopatra. It is 217 X 94 *kilometers* in diameter.[xii] This massive hunk of iron is larger than Scotland and Wales put together in area and has more PGMs than we could ever possibly use.

It is difficult to overstate the plentiful amount of resources that are available to us here on the Earth which can be derived from sources on the Moon and the asteroids. This is what we have to look forward to if we will look beyond artificial limits to growth. Over time, probably by the year 2100 if we wanted to, we could bring to an end all mining for primary metals here on the earth. With the resources of the asteroids, beginning with those that have impacted on the Moon, these dreams have the greatest potential to become reality. Asteroids have hit the Moon for billions of years, absorbing the blows, and collecting these materials, and they are waiting to be developed for our use. We go to the Moon first because it is close, three days away, rather than the two year minimum of a round trip to any of the asteroids.

Lunar Impacts

In the four billion year lifetime of the Moon, millions of asteroids, large and small, have impacted the Moon. There has been little tectonic activity to resurface the Moon, as has been the case here on the Earth. What little resurfacing there has been is starkly visible in the form of the Mare regions. The number of craters on the Moon is directly related to the age of the surface. The highland regions of the Moon are 3.8 to 4.2 billion years old and therefore have the greatest density of craters. The lowlands are 3.1 to 3.8 billion years old and have fewer craters. There should not be that much difference in the number of craters, but there is, leading to the postulation of a "heavy bombardment" period during the period just after the formation of the lunar highlands. Figure 8.4 illustrates the frequency and size distribution of highland and Mare craters:

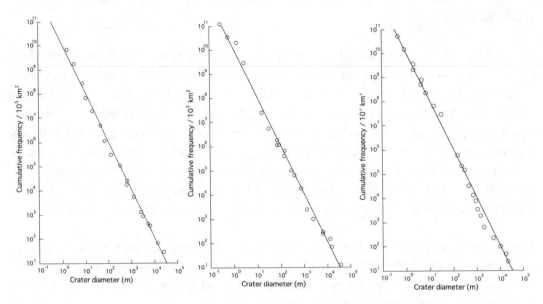

Figure 8.4: Impact Crater Frequency, Mare Tranquillitatis, Nubium, and Alphonsus

In the three charts of crater frequency vs. size, Mare Tranquillitatis has the fewest number of craters, implying the youngest age. Mare Nubium has 1.4 times more craters per unit area, and the Highlands region of Alphonsus has 2.5 times as many impacts per unit area than in Tranquillitatis, making this the oldest region of the three.[xiii] The number of craters 1 km in diameter or larger for the three regions are:

Mare Tranquillitatis	10,000 Craters per million square kilometers
Mare Nubium	14,000 Craters per million square kilometers
Alphonsus Region	25,000 Craters per million square kilometers

This data was derived from three of the early lunar impacting spacecraft, Ranger 7, 8, and 9. Dr. C. A. Cross examined the pictures from these three spacecraft and developed an inverse power law with a slope of –2 that allows for a mathematical extrapolation to allow derivation of crater frequencies and sizes outside of the resolution of the Ranger images.[xiv]

From this information some gross generalizations can be made. Since the total area of the Mare on the Moon is approximately 19% of the total surface area, and the total surface area of the Moon is approximately 38 million square kilometers, the number of craters larger than 1 kilometer in diameter is about 86,400 in the Mare regions and 845,000 in the highlands regions of the Moon. Of these impacts, 3% are M class metal asteroid impact scars. This means M class impactors make up about 2,592 impacts in the Mare regions and 25,350 in the Highland regions of the Moon. For comparison, a 1 kilometer impact is an object about the size of the Canyon Diablo impactor (Meteor Crater Arizona), which was a metal asteroid about 15 meters in diameter.[xv] It weighed nearly 100,000 metric tons and would have contained 1-10,000 kilos of PGMs.

With the total number of M class metal impactors in the range of ~28,000 objects of the same general size as the Canyon Diablo impactor, this works out to be a lot of metal. If at the absolute minimum, all of the impacts of M class objects were the same size as the Canyon Diablo impactor, the total amount of metal would be 3 billion metric tons, having 62 million kilos of PGMs (assuming 20 grams per ton average PGM concentration), 1.2 times the total amount considered commercially viable to extract on the earth by the South Africans. In truth, the amount is probably a 1,000 to 100,000 times that amount based upon the scaling law derived by Cross as shown in figure 8.4. However compelling this first brush thought experiment seems to be, the reality is a little different.

Impactor Dynamics and Survivability

It is clear that fragments of asteroids survive their high velocity travel from space to the ground on the Earth. Some pieces split or "spall" off of the main body of an impactor. Around the Barringer Meteor Crater, several tons of metal from the Canyon Diablo impactor have been recovered since its discovery over one hundred and twenty years ago. I own two meteorites from the crater and they can regularly be found for sale on Ebay (www.ebay.com) for reasonable prices. There are a huge number of metal meteorites of varying sizes and metal contents for sale on the Internet. A particularly interesting meteorite impact in Russia, the Sikhote-Alin group of objects, all nickel/iron, weighed over 70 tons. It split apart in the air and landed in hundreds of fragments, some of which can also be obtained on Ebay. Many large objects have been found in other locations, such as the 34 ton Cape York impactor, now in the Hayden Planetarium in New York, mentioned earlier and the 60 ton Hoba impactor, found near Grootfontein, shown in figure 8.5 below:[xvi]

Figure 8.5: Hoba Nickel/Iron Meteorite, Near Grootfontein, Namibia

This meteorite was found in 1920 and was not in a crater. It has been analyzed by measuring a rare radioactive isotope and it is estimated to have impacted about 80,000 years ago. The Canyon Diablo impactor is estimated to have hit about 50,000 years ago.

It should be obvious that a meteor or asteroid made basically from stainless steel would have the strongest physical materials properties. This is well borne out by analyzing meteorites of all types. A considerable amount of research has been done to determine the dynamics of the impact of asteroids and small meteors on all of the inner planets of the solar system. Below, in table 8.1, are the average approach and impact velocities of asteroids and comets on the inner planets:

Velocity in km/s	Local Planetesimals in Heliocentric Circular Orbits	Asteroids and Short Period Comets	Long Period Comets
Approach Velocity			
Mercury	2.1	19	62
Venus	5.1	15	46
Earth	5.6	14	38
Moon	5.6	14	38
Mars	2.5	8.6	31
Impact Velocity			
Mercury	4.7	20	62
Venus	11.5	18	47
Earth	12.5	18	40
Moon	6.1	14	38
Mars	5.6	10	31

Table 8.1: Approach and Impact Velocity of Asteroids & Comets on the Inner Planets[xvii]

An interesting aspect of the first column (these numbers are from NASA) is that it establishes the minimum velocity of an impacting body on any of the inner planets. Imagine for a second an object that is in almost exactly the same orbit as the earth, so the difference in velocity is very small. A way to see this is to look at a field of runners on an oval track in the Olympics. At the start, the inside runner is much farther back than the one on the outside, but only a little behind the one next to him or her. By offsetting the runners in this way it compensates for the shorter distance the runner on the inside track would otherwise run.

My research has shown that there is a discrepancy between the "official" NASA numbers and the numbers in column two and those of others in the scientific community. This discrepancy does not significantly affect the results that are discussed in this book, and I will refer to both sets of results interchangeably. The difference is usually a uniform 2 km/s. The above table refers to the average impact velocities, which is only part of the story. Figure 8.6 gives the distribution and average impact velocity of all objects, asteroids and comets on the Moon:

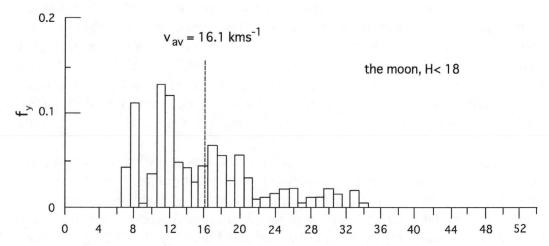

Figure 8.6: Average Impact Velocity Distribution for Impactors on the Moon

The chart above is from a paper on relative impact dynamics on Mercury versus the Moon and is the best qualitative data that I could find on this subject.[xviii] In this paper, the authors, Neukum, Ivanov, and Wagner, found that the, *"...derived projectile frequency distribution is found to be very close to the size-frequency distribution of Main-Belt asteroids...,"* which means that the NEAs that impact the Moon (and earth) are derived from the main asteroid belt. In the chart above, the vertical axis is the frequency of hits. The slowest velocity impactors in figure 8.6 correspond with the first column in table 8.1. In the chart above, most of the impactors are at the lower velocity end of the chart. The average is high due to the small number of really high velocity impactors on the right end of the chart. This is one of the dangers in relying on averages when looking at the details of a particular problem.

Looking at the chart, it is evident that the majority of the objects impact the surface of the Moon at a velocity below that of the average, while a few very fast objects skew the results toward a higher number. Velocity of impact and materials strength, as can well be imagined, governs the survivability of fragments of an asteroid, just as your survivability in a car wreck is dependent on how fast you were going and the strength of the metal of your vehicle. Also, another factor is the impact angle.

In a definitive paper by Schnabel, Pierazzo, Xue, Herzog, Masarik, Cresswell, Tada, Liu, and Fifield, in volume 285 of Science in 1999, the authors looked at the dynamics of the impact of the Canyon Diablo nickel/iron asteroid.[xix] They took a unique approach that mixed hydrocode modeling and experimental determination of the rare isotope nickel 59. Hydrocodes are numerical integration programs that compute the propagation of shock waves, velocities, strains and stresses, as a function of time and position. The nickel 59 isotopes have a half-life of 76,000 years and are created when the asteroid is in space, subject to radiation from high-energy cosmic rays. There is a relationship between the production of nickel 59 and nickel's depth inside the asteroid, governed by the shielding effect of the outer portion of the body.

By using the appropriate measurement equipment, samples of melted spherules and metal meteorite fragments from the Canyon Diablo impactor were measured. The results of the measurements showed that the solid fragments from the impactor had seven times the nickel 59 of the spherules that had melted. What this means is that the solid fragments were from the outer edges of the impactor and the once liquid spherules originated deeper in the asteroid. In their hydrocode modeling, the scientists modeled the impactor as a 15-meter spheroid impacting at either 15 km/s or 20 km/s.

In comparing their modeling to the actual amount of body estimated to have been recovered by an early researcher on the Canyon Diablo impactor (Ninninger) (4000-7500 tons of spherules and an undetermined amount of solid fragments due to removal before Barringer's ownership of the crater), they determined that the impact velocity was more likely 20 km/s. They estimated that a shell 1.5-2 meters thick, constituting ~15% of the asteroid, remained intact! For a lower velocity of 15 km/s, they estimated that the solid shell would have constituted 63% of the body that would have remained solid. Since this is not consistent with the evidence, their estimate of the impact velocity is closer to 20 km/s. This paper is online at www.dnp.fmph.uniba.sk/etext/45/text/SciCanDiab.pdf

Interestingly and importantly for the purposes of this book, at neither velocity did any part of the asteroid vaporize because the peak shock pressures did not reach the threshold for incipient vaporization of 800 giga-Pascals. The empirical evidence from the spherules is that they came from a depth of 1.3-1.6 meters beneath the surface of the impactor. This is consistent with the hydrocode modeling. These are among the definitive results of the paper that are applicable to the hypothesis concerning M class impactors on the Moon.

Let us now extrapolate this to lunar impactors of the M class metallic type. Referring back to figure 8.5, the average impact velocity of impactors on the Moon is between 14-16 km/s. This is at the lower end of the impact scale for Canyon Diablo. Therefore, using a very conservative approach, it is reasonable to expect that between 15-63% of the mass of the estimated 28,000 big impactors (those that make a crater at least 1 km in diameter) would be preserved in the general vicinity of the impact. Furthermore, there would be zero oxidation of any of these impacts due to the lack of oxygen.

Fortunately, this is not the whole story. Looking back at figure 8.5, you will see that the distribution of impact velocities extends down to a minimum of about 6.5 km/s. This is consistent with table 8.1 in the first column. It is reasonable to assume that, with lower impact velocities, the survivability of these M class metal asteroids is higher. One bit of evidence to support this comes from a paper titled, *"Meteorite Accumulations on Mars."*[xx] In this paper, the researchers used a hypervelocity gun to shoot objects of various types into material that simulates the Regolith of mars and the moon. With their system limitation of a maximum velocity of ~2 km/s and their focus on Mars, they still obtained data of interest for our lunar focus. They did fire small nickel/iron meteorite fragments into the simulated Regolith at 2 k/s and the result was deformation of the object but no further fragmentation.

Therefore, we have to separate confirmed sets of data from different researchers. We have the results of the hydrocode modeling of Pierazzo and company, coupled with the hypervelocity data from Bland and Smith. This presents a gap in the data between 2 km/s and 15 km/s that is ripe for investigation. Betty Pierazzo has indicated in personal communication that they do have data down to 10 km/s, but this gap requires more intensive study to determine the survivability of M class impactors. One can reasonably postulate that the survivability increases as velocity decreases and that the possibility exists that, at the low end of the known impact scale as shown in figure 8.5, there may be some pretty, darn big chunks of nickel/iron/cobalt/PGM objects laying on the lunar surface.

Using the absolutely conservative assumption that none of the asteroids are larger than the Canyon Diablo impactor's weight of 100,000 tons and assuming the average impact velocity of 15 km/s, we have between 450 million to 1.77 billion tons of economically recoverable nickel/iron/cobalt/PGM material in the vicinity of the impact site. Applying another conservative estimate that more realistically represents the known size-frequency curve as shown for the highland region in figure 8.3, we get nearly 1300 craters that are 10 km in size, resulting in 39 impacts of asteroids that are more like 150 meters in diameter. Taking this to the next level, we have 100 craters that are 100 km in diameter, resulting in 3 M class metal impactors that would be 1.5 km in diameter, about the size of the $20 trillion dollar 3554 Amun asteroid. This takes the resource base up to between 140-590 billion tons of localized, recoverable nickel/iron/cobalt/PGM impact material. This is on the same scale as the estimate by scientists concerning water trapped in the lunar polar region.

All of the above calculations ignore that another 1% (of the total of all impacts) of all impactors are 50-50 stony-nickel/iron bodies or that another group of asteroid impactors are Carbonaceous Chondrites that have a highly elevated PGM concentrations. Even in stony asteroids, there are a significant percentage of nickel/iron inclusions in many of them. The total amount of nickel/iron/cobalt/PGM materials on the moon is easily several times that of what is discussed in the conservative model I show here.

The only issue is that these bodies are more easily disrupted, pulverized, and splattered across the face of the moon since they are not as strongly held together as the M class metal bodies. Another factor that positively influences maximum pressures and the resulting behavior of impactors is the impact angle. The previous modeling studies referenced here assume a vertical impact angle, which is unlikely. There are also differences in the fate of the impactor depending on what type of material that they slam into. Those who slam into deep regolith versus igneous rock behave differently. The character of the lunar material also affects the explosive rebound and distribution of the rest of the body that hits the Moon. The results of this more global distribution of impact products are well known due to the Apollo program.

Global Distribution of M Class Impact Material and the Look Ahead

The Moon has a total surface area of ~38 million square kilometers. If you divide this by the total number of M class metal asteroid impacts (~28,000), you get a statistical average of approximately one 15-meter size impactor per 1357 square kilometers. That would be within an area 320 x 400 kilometers. For larger objects of ~150 meters (approximately 39 of them), this works out to one per 974 thousand square kilometers. For the largest objects of ~1.5 km, this works out to one per 12 million square kilometers. If you look at the velocity distribution curve in figure 8.5, the bias is well toward the lower velocities. This is especially true of asteroids because they all are in the lower velocity group, whose orbits evolved from the main belt of asteroids and are now not that different from the orbit of the earth. This is another limitation of the average velocity calculation in figure 8.5 since it includes the very high velocity impacts of comets. I think that the reader gets the sense that this hypothesis, while provocative to the lunar resource community, is at the very conservative end of a whole series of estimates.

Before the advent of advanced computers and the release of the classified data that forms the basis of much of the computer codes used for impact modeling (from the nuclear weapons effects world), grossly inaccurate estimates were made of impact survival rates of all types of bodies on the earth as well as the moon. In Barringer's day, the estimate was made that the Canyon Diablo impactor would have vaporized, something that is inconsistent with our knowledge today of the strength of materials and known asteroid impact velocities on various planets and moons in the inner solar system. Scientific knowledge advances in spurts sometimes and the advent of low cost, high power desktop computing is revolutionizing many fields of scientific research, including impact dynamics. Those who have held to the former view, that M class impactors vaporize, point to the dispersed nature of meteoritic nickel/iron material (between 0.1-1%) in most samples returned to the earth during the Apollo lunar missions as evidence to support their belief. You would get the same results from a resource study of the earth that took samples from six random and dispersed locations, none of them near the five major concentrations of PGMs on this planet.

I do not claim the existence of these resources in concentrated form and in the amounts estimated as a fact. This is a scientific hypothesis that is testable by remote sensing followed by ground truth studies. As a matter of fact, today, the resources of the moon are known in only the grossest form as a result of the very few missions that have been there so far. I was personally involved in the early stages of the Lunar Prospector mission and know the limitations of the instruments on that spacecraft. The instruments on the Defense Department's Clementine mission could possibly do so but the very large amount of data from that mission has not been completely analyzed and cross referenced to Prospector. Other missions have flown, but none of them have really focused in on the discovery of economically viable mining targets such as NEA impactors. Over the years, there have been many proposals for high-resolution instrumentation that have been stillborn, such as Lunar Resource Mapper, or have gone nowhere.

Up until this point, I have focused on the "why" and the "what" of the reasons for going to the moon. What has been discussed so far is only the tip of the iceberg of what is possible. To go forward from here, I will now turn attention to the "how" of getting to the moon. This will begin with a historical retrospective of over fifty years worth of architectural studies as well as what was implemented in the Apollo program. Following that will be my proposed architecture that is divergent by focusing on commercial viability rather than scientific return. What will be discussed is only one possible architecture. There are many that will work technically, but this one will be focused on the economic extraction and return of mass quantities of materials for return to the Earth. This leads to a different set of decisions than NASA has made. NASA's purpose begins with scientific discovery and only secondarily covers other concerns. This brings us full circle in the thesis of this book; that we return to the moon for the reasons that make economic sense, not just for science or for some intangibles. The intangibles will come from the activity and science will be increased dramatically by the economic activity proposed.

Next, we will turn to ways to test our hypothesis. Any hypothesis that does not have a way to be tested will forever remain speculative. Happily, this is not the case with metal impactors on the lunar surface, as we shall see.

*   *   *   *   *

[i] Wertime T and J. Muhly, *The Coming of the Age of Iron*, Yale University Press, New Haven and London, 1980. P. 71-73.
[ii] Ibid, P. 512
[iii] J. Lewis, et al, editors, *Resources of Near Earth Space*, University of Arizona Press, Tucson and London, 1993, P. 533
[iv] http://www.auroraplatinum.com/s/Projects.asp?PropertyInfoID=743&View=All
[v] http://www.unb.ca/passc/ImpactDatabase/images/sudbury.htm
[vi] http://www.unb.ca/passc/ImpactDatabase/images/vredefort.htm
[vii] W.E. Elston, *The Proterozoic Bushveld Complex, South Africa: Plume, Astrobleme, or Both?*, Department of Planetary Sciences, University of New Mexico
[viii] http://news.bbc.co.uk/2/hi/science/nature/3281611.stm
[ix] *Resources of Near Earth Space*, P. 529
[x] Mining the Sky, P. 112
[xi] Ibid
[xii] http://www.acseal.freeserve.co.uk/asteroid.htm
[xiii] C.A. Cross, *The Size Distribution of Lunar Craters*, Monographs of the Royal Astronomical Society, 1966, P. 245-252
[xiv] Ibid
[xv] http://www.barringercrater.com/
[xvi] http://www.canadianrockhound.com/2000/02/cr0004212_meteorite.html
[xvii] http://cmex-www.arc.nasa.gov/CMEX/data/SiteCat/sitecat2/crater.htm
[xviii] G. Neukum et al, *Crater Production and Cratering Chronology for Mercury, Mercury: Space Environment*, Surface, and Interior, 2001 (8027.pdf)
[xix] C. Schnabel, et al, *Shock Melting of the Canyon Diablo Impactor: Constraints from Nickel-59 Contents and Numerical Modeling*, Science, Volume 285, 1999
[xx] P. Bland, T. Smith, Meteorite Accumulations on Mars, Icarus 144, P. 21-26, 2000, online at www.idealibary.com

Chapter 9:

Prospecting The Moon

Before any real development of the Moon takes place, it is necessary to execute a suite of missions that, while they may have great scientific value, are principally for the purpose of prospecting for resources on the Moon. To someone who is not initiated in the study of the Moon, it is reasonable to expect that these missions took place decades ago during our first campaign to go to the Moon during the Apollo era. The sad fact is that the missions flown during the Apollo era, and even those since then, have created an incomplete amount of information for the geological resource community to conduct a complete resource study, especially if the goal is to find nickel/iron impactors. I will begin here by describing the earlier missions and their purpose. Later in chapter 14 I will go into detail concerning these prospecting missions as part of a return-to-the-Moon commercially oriented architecture.

The History of Lunar Science Missions

For centuries, scientists, amateurs, and lovers have gazed into the heavens at our nearest neighbor. To me the Moon is a beautiful object and even with a modest telescope you can see the shadows of mountains, really detailed craters and the dark grey lunar mare. Those of us who claim expertise in lunar research can scarcely believe how ignorant we all were at the dawn of the space age in 1957. An example is that even up into the 1960s, a debate raged about whether or not the craters on the Moon were the result of impacts.

This is not so impossible to believe when you read books from that era, such as, "The System of Minor Planets," by Gunter Roth.[i] At that time, we knew about the asteroid belt and knew that there were at least a few thousand bodies between the orbits of Mars and Jupiter. Both the famous Gerald Kuiper and the South African Baade estimated the total at no more than 4,000 objects with a brightness of magnitude 17 or better (this is for a stellar magnitude of brightness of objects, this is generally far fainter than can be seen by the naked eye). However, at that time, only a few NEAs were identified. With the information that was available to the scientific community at the time regarding what they felt was a paucity of asteroids, it was reasonable to postulate non-impact processes for lunar craters. We now know that there are literally tens of millions of main belt asteroids, and hundreds of thousands to millions of NEAs. We have extended the range of our observation all the way to the 29th magnitude by the Hubble Space Telescope and have discovered rocks as small as a few meters across. With this knowledge and a lot more data from the Apollo program, asteroid impacts on the Moon are an accepted fact.

First Efforts

The first spacecraft from the Earth to reach the vicinity of the Moon was the Russian Luna 1 spacecraft in January 1959. It missed the Moon, but its successor, Luna 2, did manage to hit the lunar surface several years before a comparable American effort. The Russian Luna 3 spacecraft nabbed a historic first by taking the first picture of the far side of the Moon in October of 1959. Figure 9.1 is a picture of the first two types of Russian lunar spacecraft:

Figure 9.1: Early Russian Lunar Spacecraft

When these early missions by the Russians and the pictures taken were revealed to the world, the most interesting thing shown was that the far side of the Moon does not look anything like the near side. While the near side of the Moon has very prominent "seas" or mare, the far side has very little of this texture. The far side is mostly highlands, an older surface, with far more craters than we can see on the Earth facing side. The Moon always faces the Earth because the Moon is "lumpy," or asymmetrical and the gravity of the Earth eventually slowed the rotation of the Moon to lock with its orbit around the Earth. The greater number of craters in the highlands versus the mare led to the theory of an "early bombardment" period. It is thought that far more asteroids hit in the early years of the Moon, from 3.8-4.2 billion years ago, than hit later, after some really big impactors hit the Earth-facing side that left a molten surface, creating the mares that we see today.

Figure 9.2: Ranger Series of Spacecraft (Picture Courtesy NASA)

The first successful American flight series began with Ranger 7 (the first six failed) in July of 1964 and ended with Ranger 9 in March of 1965.[ii] Figure 9.2 is a picture of a Ranger series spacecraft:

The Ranger spacecraft carried a video camera that sent back pictures until the time that the spacecraft impacted on the lunar surface. There was no attempt to go into orbit since the launch vehicle technology of that time and the in-space navigation capability of the U.S. was not up to the task for such a mission. The three spacecraft returned over 17,000 images with resolutions varying from kilometers to well less than a meter just before impact. There are some excellent resources on the Internet for detailed information on these spacecraft from the University of Arizona and from the National Space Science Data Center (NSSDC). A search term of [Ranger Lunar Spacecraft] will return a wealth of information.

Landers

Just after the time that the Ranger series of spacecraft was wrapping up, the Russians again leapt ahead of their American rivals with new spacecraft in the Luna series that actually achieved the first soft landing on the Moon in early February of 1966. This followed both the U.S. and Soviets blowing up, crashing or otherwise destroying spacecraft before getting it right. Figure 9.3 shows the Luna spacecraft in its deployed configuration next to a photo of the first successful American lander series, Surveyor:

Figure 9.3: Left Luna 6 (Russia),
Right Surveyor III with Apollo 12 in the Background

The Luna Spacecraft shown on the left pioneered the air bag landing system later adopted by the U.S. in landing the 1997 Mars Pathfinder and the Spirit and Opportunity rovers in early 2004. The photo on the right shows the Apollo 12 crew visiting the landing site of the third Surveyor spacecraft. Both of these spacecraft series were used more as engineering tests than science. They both took pictures and the Surveyor had a small robotic arm that scooped up some regolith. The last three Surveyors also carried an alpha spectrometer, an ancestor of one of the instruments taken to Mars on Spirit and Opportunity. All of the Surveyors were successful and took tens of thousands of pictures of their surroundings. The principal finding of the Surveyors was that the lunar surface was not covered with a deep quicksand-like regolith as some scientist had feared. One of the spacecraft actually took off again and moved a few meters away as a further test of the strength of the lunar surface regolith. Unfortunately, they all landed in relatively boring areas that were precursors to the Apollo landing sites.

The Russians later soft-landed Luna 16 in September of 1970, it used a drill to collect a 101 gram regolith sample which was sealed and returned to Russia. This was another first for the Soviet space program. Figure 9.4 is an image of the Luna 16 lander:

Figure 9.4: Luna 16 Lander/Sample Return System

The Russian lunar program continued with more lander/sample return missions and even carried the first lunar rovers on two separate missions. The Lunokhod 1 and 2 rovers were impressive systems that survived for several months in an environment far harsher than that endured by Spirit and Opportunity today on Mars. Lunokhod 1 traveled more than 10 kilometers (6.3 miles), taking pictures and conducting scientific experiments. Lunokhod 2 was even more impressive, traveling over 37 kilometers (22 miles) with multiple scientific experiments and an imaging system that sent back over 80,000 images. In comparison, Spirit and Opportunity together have traveled less than ten percent of the distance traveled by Lunokhod 2 alone. This is even more impressive considering that these missions happened three decades ago.

Orbiters

Luna Series

In the 1960s, both the Soviet and American space programs placed mapping systems into lunar orbit. The Russians sent Luna's 10, 11, 12, and 14 into lunar orbit in 1966 through 1968. They carried gamma and x-ray instruments and also an imaging system on Luna 12. Little is known of the images and no information was available from the NSSDC on the other instruments.[iii] The Russians sent a later series of Luna's (19 and 22) that carried a gamma ray spectrometer, a solar wind experiment and an imaging system. The particle experiment data is now available from the NSSDC for download and some or all of the images can be ordered, though the web site says that the images have been degraded due to time. Figure 9.5 shows both series of Russian Lunar Orbiters:

Figure 9.5: Luna 11 Series Orbiters (Left) and Luna 19 Series (Right)

Lunar Orbiter

The United States began their lunar orbiting campaign in August of 1966 with five identical photo recon orbiters.[iv] The first three orbiters mapped a swath of the lunar surface in the equatorial plane to select landing sites for the Apollo missions. Figure 9.6 is an image of the Boeing built spacecraft:

Figure 9.6: Lunar Orbiter Spacecraft (Picture Courtesy NASA)

The last two spacecraft created a global map of the Moon at resolutions from 60-140 meters. There is a lot of information available online associated with this mission. Lunar Orbiter mapped about 95% of the total lunar surface, although some of the images are of poor quality. The images were originally generated on film on board the spacecraft that actually had its own film processing system. The pictures were then digitized and transmitted back to the Earth. These images were reconstructed and stored on tape. Theoretically, images of equal quality to the originals can be duplicated today. An interesting feature of some of the pictures is, that if there was a flaw in the film processing system on the spacecraft, it gets translated and shows up on the final image. Figure 9.7 is the photo overlay of the images on the near side lunar surface:

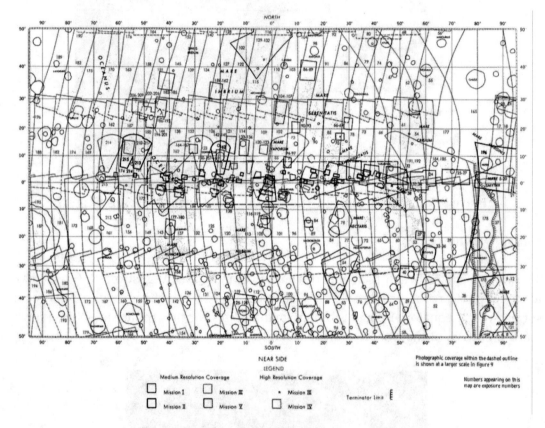

Figure 9.7: Lunar Orbiter Photo Overlay for the Five Mission (Lunar Near Side)
(Picture Courtesy USGS)

These missions comprise the first complete data set of mixed high and medium resolution images. Unfortunately, the highest resolution images are not available online yet. Following are some of the best websites for the Lunar Orbiter data:

http://www.lpi.usra.edu/research/lunar_orbiter/

This website has the data catalog of all Lunar Orbiter images that can be browsed by several different methods. This is the first site to go to for images to begin any research on mapping the Moon. The caveat is that this data is at a resolution considerably less than the true resolution of the images, which limits their value. However, they do have a companion dataset that shows the proper names of features and so is a great learning tool for identifying lunar features.

http://astrogeology.usgs.gov/Projects/LunarOrbiterDigitization/

This website is from the U.S. Geological Survey and is a project in progress to digitize very high-resolution images from first generation negatives as close to the original data as possible. These images are of a higher resolution than anything that I have seen anywhere publicly available.

http://www.lpi.usra.edu/research/cla/maps/list/

This website has pictures from ground based astronomy that is suitable for determining what images you want at higher resolution from the orbiting missions.

All in all, the Lunar Orbiter missions were very productive for the needs of the Apollo program, but were really not good for a more global view. Figure 9.8 is an example of a cropped, high-resolution image from the Lunar Orbiter III frame 200 H2:

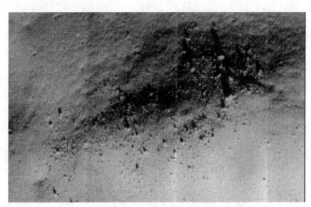

Figure 9.8: Lunar Orbiter III frame 200 H2 (Picture Courtesy USGS)

This picture is very interesting and suggests what is possible with a global map at one meter resolution. Individual boulders are clearly visible in the image.

Other 1960s Orbiters

During the Apollo 15 and 16 missions, the Command Service Module (CSM) ejected a small spacecraft that carried a magnetometer, an S band transponder experiment and a charged particle experiment.[v] This is an important spacecraft for studying resources because measurements of magnetic field strengths can help to determine the location of large M class impact products or a large amount of closely distributed material.

On Apollo 15, 16, and 17, the CSMs also carried several experiments that obtained very good science and were good engineering tests of more capable instruments carried by later spacecraft.[vi] These included:

Laser Altimeter	10 Meter Resolution
Gamma Ray Spectrometer	Iron, Titanium, and other Atomic Spectra
Alpha Spectrometer	Radioactive Element Information
X-Ray Spectrometer	Magnesium, Aluminum, and Silicon Detection
Bistatic Radar	Electrical Properties and Roughness of the Surface
Mass Spectrometer	Gas Composition and Distribution
S Band Transponder	Gravity Mapping

All of these instruments obtained good data, but only for the equatorial regions of the Moon that Apollo spacecraft flew over. Figure 9.9 is an example of the laser altimeter data:

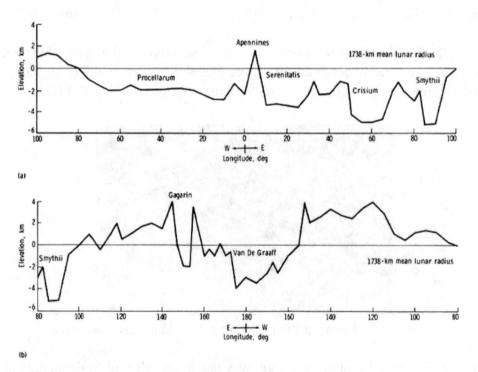

Figure 9.9: Apollo Laser Altimeter Data (Courtesy Lunar Planetary Laboratory)

All in all the Lunar Orbiter missions were very productive for the Apollo program needs but were really not good for the purpose of resource prospecting. It is hard to criticize the NASA folks in the 60s since their goal was to send people to the Moon and return them to Earth safely, which they did do, but to not take the next step and at least do global maps of elemental abundances, minerals, topography and visible and infrared images, showed a very disturbing lack of foresight, if not by NASA then the congress that funded them. For example, the Gamma Ray spectrometer data only covered 19% of the lunar surface. The X-ray spectrometer data only covered 9% of the lunar surface. The resolution of the Gamma Ray data at 100 kilometers resolution and the X-ray data at 15 km was inadequate for the detailed remote sensing of resources. Laser altimeter data such as that shown in figure 9.9 was only taken for a few orbits.

Clearly, the Apollo era data is very limited, both in terms of quantity and quality. The restrictions on instrument design are excusable, but considering the staggering amount of taxpayer dollars spent on Apollo, one would have hoped that a dedicated spacecraft for a resource related remote sensing mission would have been flown. During the Apollo program, no further orbiters were sent to the Moon nor were the remaining three Saturn V launch vehicles funded for lunar missions. This was decided after the money was already spent for their construction. This action by the U.S. government defies reason in their decision to cancel the final missions much less the whole program. NASA's stated reason for cancellation was concern about crew safety. Considering the problems that were confronting the world in the early 70s and the potential for space to contribute to solving these problems, the decision to terminate most human space exploration will be debated for centuries.

Modern Lunar Orbiting Missions

From 1972 until 1990, the United States sent no spacecraft at all to the Moon. The Russians continued for a short while with their Zond and Luna series, but stopped in 1976 after a successful program of robotic orbiters, landers, rovers, and sample return missions. From 1990 until today we have sent two orbiters and two opportune flybys by Galileo. The Russians have sent no more missions and the Japanese have sent two simple spacecraft. The Europeans have entered the fray and their new generation SMART-1 spacecraft has begun long-range science operations as of May 2004 with an expected 18 month mission beginning in 2005 for operations in a low lunar orbit.

In comparison, between 1972 and 1990 the U.S. sent two robotic landers and three orbiters to Mars. Since 1990, the U.S. has sent three orbiters to Mars, two of which are still in operation, and five landers/rovers. Three of the rovers were successful. Two of them (Spirit and Opportunity) are still functioning as of May 2004. Several more missions are planned between now and 2009, when so many spacecraft will be in Mars orbit that a communications relay satellite will be sent just to coordinate science data flow.

To date, none of the spacecraft sent to the Moon were meant specifically for a detailed resource analysis. The closest to a "resource" mapper that actually flew was the appropriately named Lunar Prospector, originally a private mission that was eventually funded by NASA as the first of their "Faster, Cheaper, Better," Discovery class missions.

Galileo

The Galileo spacecraft was the first modern spacecraft that obtained remote sensing data of the Moon. However, the Moon was not its original target. It was actually intended to orbit Jupiter for several years. Due to limitations related to its propulsion system, Galileo had to perform multiple flybys of inner solar system planets, specifically Venus and the Earth, to obtain enough orbital energy to make it to Jupiter. First in 1990, and second in 1992, Galileo turned its sophisticated multibillion-dollar instrument set toward the Moon from fairly close range. The first flyby, on December 8, 1990, was a long distance flyby at a range of 624,858 km (390,000 miles). The closest approach was to Mare Orientale. Galileo mapped the Moon from this range with its solid-state imager and its near infrared mapping spectrometer.[vii] Almost exactly two years later, on December 7, 1992, Galileo flew by the Moon for the second time. This time, in addition to visible light images with decent resolution, other instruments were turned on to examine the Moon for mineral concentrations.[viii] This did not significantly add to the image quality because the best resolution was only 1.1 km, but the information gathered by the other instruments added to our knowledge of the elemental composition of the Moon. Plate 5 shows a fascinating false color image of the lunar surface that does give a coarse indication of lunar resource potential:

Titanium rich soils of Mare Tranquillitatis are shown in blue on the left. Lower concentrations of titanium are shown in orange. The dark red color indicates areas that are low in both iron and titanium. The purple patches are from volcanic activity. While the Galileo flybys were brief, they did show the potential of missions dedicated to the Moon carrying advanced instruments.

Clementine

The first lunar orbiting spacecraft specifically intended for that destination was a mission conceived at the U.S. Defense Department's Strategic Defense Initiative Organization (SDIO) (Now the Missile Defense Agency) in 1989. Named Clementine or Deep Space Program Science Experiment DSPSE, this innovative technology development mission was designed to test a wide range of advanced sensors in a realistic operational environment.[ix]

Rather than operate in Earth orbit where the operation of a strategic defense satellite might generate opposition no matter how benign the mission, Col. (now Brigadier General Retired) Simon "Pete" Worden, Col. Pedro "Pete" Rustan and Stuart Nozette, decided to send the spacecraft to the Moon where it would gather significant remote sensing data while testing out instruments of military utility. Col. Worden was an astronomer by training and Col. Rustan was a maverick by nature so this mission was approved. Clementine's second mission goal was to leave lunar orbit and proceed to do a flyby of the NEA 1620 Geographos. This mission goal failed due to a software glitch that caused the depletion of attitude control fuel after the lunar mission was completed. Figure 9.10 is a computer rendering of the deployed Clementine spacecraft:

Figure 9.10: Clementine (Deep Space Program Science Experiment DSPE)

Clementine was an inexpensive mission compared with similar NASA efforts. For a total cost of $80 million dollars, Clementine was designed to generate an impressive amount of data at the Moon as well as at 1620 Geographos. Clementine carried the following science payload:

Ultraviolet/Visible CCD Camera (UV/VIS)

This instrument was used to obtain a visible light global map of the lunar surface to a resolution of between 100-325 meters per orbit. The images were taken in the visible light range of 250-1000 nanometers (nm).[x] This was roughly comparable to Lunar Orbiter IV and V in resolution and provided an updated map that helped to compliment the Lunar Orbiter data and fill in a few of the gaps left by those missions. However, the Clementine imaging system was far superior because it carried several narrow band filters (415 [40nm], 750 [10 nm], 900 [30nm], 950 [30nm], and 1000 [30nm]), as well as a 400-950 nm broadband filter.

The visible light imaging system obtained over 400,000 images, covering 100% of the lunar surface. The short and mid wavelength filters at 450 and 750 nanometers were used to create a map of the elemental distribution of titanium across the lunar surface. This is shown in Plate 6. The mid wavelength filters at 750 and 950 nanometers were used to create an iron distribution map shown in Plate 8. These maps were verified against the ground truth studies from the Apollo lunar samples returned during those missions. Other maps of great interest include albedo maps of the Moon that give relative lighting across the entire lunar surface. This helps scientists to determine regions of total lunar darkness for sites of possible water on the Moon.

These are very valuable maps of the overall distribution of resources across a wide swath of the Moon. A second extremely valuable use of the imaging system was to build a movie of the lighting conditions at the South and North poles of the Moon. This helped to narrow the region of possible permanent darkness and help focus scientific efforts to locate water in the areas that never get sun, called cold traps. Another very valuable contribution of Clementine for visible light imaging is that an almost 30 year gap between Lunar Orbiter and Clementine allows for a nice comparative analysis of the lunar surface over that period of time. Plates 9 & 10 show the North and South Poles built into a mosaic from Clementine data. Plates 11 & 12 show Iron concentrations at the Apollo 16 landing site and in the Oriental Basin. Plate 12 has an interesting iron anomaly, that in this instance is from volcanism. However, this image is how a high spectral resolution PGM metallic impacter might look. Images and movies of this are at www.lpl.arizona.edu at the University of Arizona's Lunar and Planetary Laboratory. A lunar image browser of the UV/VIS images is located at the Naval Research Laboratory (NRL) website at http://www.cmf.nrl.navy.mil/clementine/clib/.

Near-Infrared Camera (NIR)

This camera took images of the Moon in the near infrared wavelengths between 1100-2800 nanometers in six different sub-bands. This is a really nice imager with a field of view of 5.6 X 5.6 degrees, obtaining pictures about 40 kilometers wide.[xi] This instrument also obtained a global map of the Moon at several wavelengths not obtained prior to the Clementine mission. The NIR imager obtained infrared images at 1100, 1250, 1500, 2000, 2600, and 2800 nanometers, all with 60 nanometer bandwidths. Several hundred thousand images were generated, creating a global dataset at these wavelengths. The maximum resolution of these images was between 150-500 meters. The huge amount of data from this instrument was calibrated and released to the NASA planetary data system in 2000.[xii] This dataset is not online as of this time, but it is available through the NASA planetary data system. This data is directly applicable to the search for nickel/iron meteorites on the Moon because the 100 meter resolution would seem to be right on the edge of that required to be able to identify the different infrared signatures of large nickel/iron fragments or a large debris field.

Long Wave Infrared Camera (LWIR)

Clementine carried a thermal infrared imaging camera that used a Mercury/Cadmium/- Telluride focal-plane array.[xiii] This camera was designed to image features on the Moon during local lunar night and to allow the measurement of the thermal properties of the regolith and other material on the lunar surface. The field of view for this instrument was 1 X 1 degree. The resolution of the instrument was about 55-136 meters per pixel. The camera imaged approximately 0.4% of the total lunar surface, mostly near the poles (85-90 degrees). A few other interesting areas were imaged as well. Even with this limited coverage, the instrument generated 220,000 images! This would have been an incredible instrument to use closer to the lunar equator just after sunset to obtain high-resolution infrared data that would have had a good probability of finding large nickel/iron bodies or strewn debris fields of the same. The thermal radiation characteristics of nickel/iron bodies should have shown up fairly well in these circumstances.

High-Resolution Camera (HIRES)

The HIRES camera on Clementine consisted of a 288 X 384 pixel CCD with five narrow band filters at (415 [40nm], 560 [10], 650 [10 nm], 750 [20 nm],) nanometers. This was a very narrow (0.3 X 0.4) field of view camera with a maximum resolution of 7-20 meters. The areas of the Moon mapped by the HIRES instrument extends from 80 degrees S and 80 degrees north with a resolution of 20 m/pixel and 30 m/pixel resolution at the poles.[xiv] It is unclear from the NSSDC website whether or not this is a global map. The image data from the HIRES instrument is available from the NASA Planetary Data System, but only limited amounts are available online.

LIDAR

The Clementine spacecraft carried a laser radar, or LIDAR, that used a laser beam in a mode similar to a laser altimeter.[xv] The resolution was 40 meters and a pulse was sent out once each second over a period of 45 minutes. This aided scientists like Paul Spudis to generate a global topographic map of fair resolution.

Bistatic Radar Experiment

With regard to concentrated resources, the most interesting experiment was the Bistatic Radar Experiment. This experiment used the 2.273 GHz radio transmitter to transmit a signal down into the permanently shadowed regions of the lunar north and south polar regions. This experiment's goal was to have Earth bound receivers analyze the signals to look for data that might indicate the presence of water in these cold polar traps. The results are controversial and, over the years, different researchers have come to different conclusions about whether or not the Clementine data actually conclusively indicates water in these regions.

Clementine Summary

The Clementine mission was an incredible advance for lunar science and resource prospecting. The mission was accomplished for a very moderate price, especially considering the quantity and quality of data returned. Data reduction and analysis of Clementine's instrumental data is still an active subject of research. Preliminary studies of the data have the potential of discovering NEA M class impactors. However, for all of its value, Clementine is still not quite what is necessary for this study and its inconclusive results for water have just whetted the appetites of resource recovery advocates. The other orbital mission flown during the 1990's, this one by NASA, was the Lunar Prospector mission, whose purpose was to "prospect" for lunar resources.

Lunar Prospector

The Lunar Prospector Project was begun as the result of the frustration of a lot of space advocates concerning NASA's lack of interest in the Moon. The name "Lunar Prospector" was specifically chosen by those involved (this author included) to make the point that we wanted to prospect for resources on the Moon. The project began with a great deal of enthusiasm and volunteers, and a small amount of funding by Gerard K O'Neill's Space Studies Institute. Figure 9.11 is a rendering of Lunar Prospector in orbit around the Moon:

Figure 9.11: Lunar Prospector in Lunar Orbit (Picture Courtesy NASA)

Over time, the project moved through many fits and starts. It was the perseverance of the Principal Investigator, Allen Binder, that eventually moved the project to the point where it was accepted by NASA as the first mission of its "Faster, Cheaper, Better" philosophy. This was NASA's way of copying the success of SDIO's Clementine mission. Dr. Binder was able to build and fly Lunar Prospector for $63 million dollars, a testament to his goal of building a low cost mission and also to the sweat and labor of all of those who volunteered their time and energy to conceive of and support the early development of the spacecraft.

Prospector was unique in its selection of instruments. The principle instrument was a Gamma Ray Spectrometer (GRS) that would finish the work begun by the Apollo instrument flown on Apollo 15 and 16. Following is a description of the Prospector instruments:

Gamma Ray Spectrometer

This was the instrument that everyone agreed had to fly in the early days of the project. Its goal was to map the global elemental abundance of iron, titanium, thorium, potassium, oxygen, silicon, aluminum, calcium, uranium, and magnesium. The instrument, built by the Los Alamos National Laboratory, was designed to measure gamma rays that originate on the lunar surface by the impact of high-energy cosmic rays on the elements listed. Each gamma ray generated has a specific energy that can be measured by the instrument. The energies of the gamma rays indicate the relative abundance of the elements on the lunar surface to a depth of approximately one meter.

A borated plastic shield wrapped around the bismuth germanate crystal of the GRS was used to detect epithermal neutrons reflected from the surface by hydrogen atoms, presumed to measure the abundance of water in the polar cold traps. This instrument has to take data over a long period of time in order to build up a statistically meaningful amount of data. The original Prospector mission was six months and was extended for another six months to accumulate more data.

The most exciting result of the operation of the GRS was the successful detection of the desired epithermal neutrons over the north and south poles of the Moon. Unfortunately, the resolution of the instrument (~100 kilometers) did not give clear indications of exactly where the hydrogen was specifically located. The precise location of the hydrogen was inferred by the complimentary Clementine data that located the cold traps with its imaging system. Plate 7 shows the presumed location of the hydrogen-bearing regions on the Moon. Figure 9.12 is a chart showing the dip in readings associated with the hydrogen-bearing regions of the lunar poles:

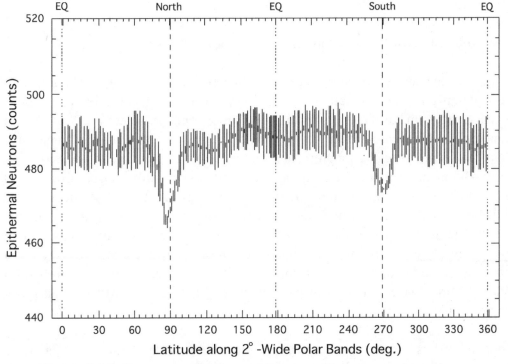

Figure 9.12: Prospector Epithermal Neutron Count (Picture from Paul Lucy)

The most striking aspect of the figure 9.12 graph is the suggestion that there is more water at the north pole than the south.[xvi] The color map in plate 7 shows this as well, but not as convincingly. The presentation by lunar scientist Paul Lucy (reference XVI) is actually a very funny and informative document concerning the real meaning of the Lunar Prospector GRS data. Other researchers, including Binder, Nozette and too many others to mention here, variously estimate the amount of water to be somewhere between 10-1,000 million tons. This is quite a lot of variability! Almost all of the other researchers equate the existence of hydrogen with water, but there is no evidence that this is the case. Dr. Lucy makes the astute observation that, if the water is implanted by comets, asteroids, or interstellar clouds that waft through the solar system, the hydrogen is just as likely to be in the form of hydrocarbons. This is not a big deal for those of us who want to use this resource, except that it is much easier to turn ice, rather than methane or tar, into high-power rocket fuel.

The Significance of Water on the Moon

The importance of water as a resource exported to the Earth from the Moon is zero. The importance of water on the Moon for local uses is incredibly large. To ship water from the Earth to the Moon for rocket fuel or any other uses has been estimated to cost $35,000 per pound. If it is already there on the Moon, it makes the logistics of a lunar economy so much easier. It is not definite that the water is there in a way that makes it an economical resource. Before we return to the Moon, it is vital that we answer these important questions by sending spacecraft with better instruments to improve the quality of data for our search for resources.

The Problem With the Missions that Have Gone Before

As good as the data has been from previous lunar orbiting missions, including Apollo, we have only begun to create a real, global scale lunar mapping of resources. The Lunar Orbiter image data was at high-resolution only in select areas of the lunar equator. The Clementine data, while good and complimentary to the Lunar Orbiter Data, still is not at high enough resolution to be able to plan landing missions at the all-important lunar poles. Clementine's multiple wavelength images, from the NIR and LWIR instruments, need to be better if we are going to use them to detect nickel/iron M class fragments in the lunar regolith. The Lunar Prospector data, while it is the first global map of the distribution of ten important elements, lacks resolution and sensitivity compared with the instrument on the current Mars Odyssey mission that found a huge amount of water on Mars. It is my estimate that one to two other lunar orbiting spacecraft need to be sent to the Moon to improve the datasets that already exist.

None of this is meant to denigrate the incredible work of both Clementine and Prospector. If we had not done these missions and retrieved the data that they took, lunar science and our efforts to plan for the use of lunar resources would be even farther behind where it is today. The really good news is that the instruments for these studies and the capabilities of spacecraft to handle large amounts of data have dramatically improved since these two missions. One of Clementine's unheralded accomplishments was its use of state of the art computers (for 1992), memory and on-board software, which completely disproved some of the conservative assumptions of NASA and the Defense Department about what computer hardware could be flown. However, the fact remains that we have made quantum leaps in computer power, communications technology and software for data collection that will make all other spacecraft's data obsolete.

Current and Future Lunar Missions

For a taste of what is possible, let's look at Europe's first entry into the lunar derby: ESA's fantastic Small Mission for Advanced Research and Technology or SMART-1 spacecraft. ESA chose the Moon as its first target.[xvii] Figure 9.13 is a rendering of SMART-1 on the way to the Moon:

Figure 9.13: SMART-1 To the Moon

SMART-1 is the first of a new generation of spacecraft from Europe using an innovative ion propulsion system from Russia. This ion propulsion system, called a Hall Effect Thruster, accelerates xenon atoms, stripped of some of their electrons, to a high velocity, which provides thrust for the spacecraft. The Hall thruster uses large amounts of electricity (the SMART-1 thruster uses 1350 watts of power), provided by solar arrays, to accelerate the xenon atoms. This technology has flown for decades on Russian spacecraft, but they have not been available in the west until the fall of the Soviet Union. Hall thrusters are very efficient compared to chemical propulsion. They are so efficient that SMART-1 was able to launch as a secondary payload along with two commercial satellites going to geosynchronous orbit. This saved a lot of money and gave a good starting orbit to SMART-1. The total cost of the SMART-1 spacecraft was 110 Euro, about 130 million dollars, including launch and operations.

Actually, SMART-1 is a technology development mission that is being used to test and flight qualify several different technologies for future ESA missions. As of late May 2004, the SMART-1 Hall ion thruster has accumulated over 2000 hours of operation. Several thousand more hours are required to get the spacecraft from its current orbit to the Moon. Figure 9.14 shows the SMART-1 ion thruster in its test chamber at ESTEC in the Netherlands:

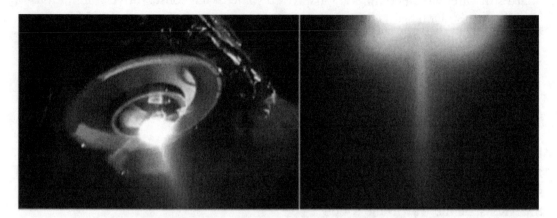

Figure 9.14: SMART-1 Hall Thruster System in Test

This system may be slower than chemical propulsion, but there are huge advantages to this approach. The spacecraft is smaller, lighter and more capable than a chemically fueled system. It may even be possible to "fly" the SMART-1 spacecraft in an extremely low orbit (this is under consideration at ESA) that will allow for very high-resolution imaging of the lunar surface in the infrared, visible and x-ray spectrum. Following is a brief description of its instruments.

AMIE (High Resolution Micro-Camera)

The AMIE imaging system is a marvel in microminiaturization. It weighs 2.1 kilograms and draws 9 watts of power. The CCD imager is an off-the-shelf commercial item with a 1024 X 1024 pixel matrix. It supports three separate filters at 750, 847, and 960 nanometers. It has already successfully imaged the Moon and is expected to be able to image the lunar surface at a resolution of better than 100 meters per pixel and as good as 40 meters in the southern hemisphere of the Moon. This will compliment, but not replace the Clementine HIRES images. One benefit will be long-term observations of the lunar poles that will allow a final definitive determination of whether or not there are areas at the lunar pole that are always in the sunlight. This is extremely important to future selection of lunar base sites since it will make possible a reliable source of energy. Depending on how long the mission lasts, stereo views of the lunar surface under various lighting conditions will be obtained.

SIR

The Solid State Infrared Imager will take images from 930 to 2400 nanometers. This is roughly comparable to the spectral range of the NIR imager on Clementine. However, instead of six filters, the SIR will have 256 channels across the entire spectral range. This is a considerable improvement in technology from Clementine and will enable the discrimination of several different types of minerals of interest to science and to resource extraction efforts. The SIR will also look into the deep shadowed regions to determine the type of hydrogen resources in the cold traps. The resolution on the images will be 300 meters, which is less than the NIR but will make up for it in spectral resolution.

D-CIXS X-ray Spectrometer

The D-CIXS spectrometer is a demonstration device for future missions. It carries a bank of 24 SCD detectors with an energy range from 0.5 to 10,000 electron volts. It has a field of view of 12 X 32 degrees. This will be the first instrument to map the Moon in the X-ray bands for aluminum, magnesium and silicon. Its theoretical maximum resolution is 40 km, but in a recent paper by the SMART-1 scientific team, it is stated that the realistic resolution will be approximately 100 kilometers except in the polar regions.[xviii] This could improve if SMART-1 flies in a lower altitude orbit as the mission progresses.

SMART-1 Importance to Lunar Science and Resources

The SMART-1 mission will obtain critical information that will be used by U.S. and European planners for potential sites for future lunar bases. This same information will be of use for those who seek to use the resources of the Moon for commercial enterprise. The data from SMART-1 will complement and supplement the data from Clementine, Lunar Prospector, Apollo, and Lunar Orbiter. SMART-1 will help to improve our knowledge of the amount of hydrogen there is at the lunar north and south poles, which is also very important for planners of future landing missions for science, logistics, and resource utilization.

SMART-1 is also helping to flight qualify a new generation of advanced technology for lunar and other missions with its innovative Hall thruster system. There are some limitations of SMART-1 related to its use for a potential prospecting mission. SMART-1 will not provide the types of high-resolution data required to pinpoint lunar resources of nickel/iron M class impact fragments. For this, a new mission is required.

Future Missions

Whether or not the United States moves forward with lunar research, we are entering into the most prolific phase of lunar exploration since the death of the Apollo program. Going beyond the current SMART-1 mission by ESA, the Japanese space agency JAXA has two missions in the advanced stage of construction. The first of these missions is the Lunar-A mission, currently scheduled to launch late in 2004 or early 2005. A second mission by JAXA is the Selene mission to the Moon, scheduled for launch sometime in 2005.

Both of these missions are orbiters with the additional bonus that the Lunar-A mission will carry two penetrators that will impact the lunar surface at a velocity of 250-300 meters per second. Both of these missions are capable spacecraft that will continue to supplement the data obtained by Clementine and Lunar Prospector. Figure 9.15 shows the instruments, their spectral bands and their resolution on the lunar surface for Selene and Lunar-A:

Project	SELENE					
Country	Japan					
Launch	2005					
Spacecraft altitude of observation phase	70 -130km					
Spacecraft inclination	90deg.					
Spacecraft control	3-axes-stabilized					
Mission life	1 year <					
Instrument	Terrain Camera (TC) - 1, 2	Multi band Imager (MI) - VIS	Multi band Imager (MI) - NIR	Spectral Profiler (SP) - VIS	Spectral Profiler (SP) - NIR1	Spectral Profiler (SP) - NIR2
Field of view (FOV)	19.3 deg. (max. 22.4 deg.)	11 deg.	11 deg.	0.29 deg. x 0.33 deg.		
Instantaneous FOV	20"	40"	123"	-		
Spatial resolution	10m (@ 100km)	20m (@ 100km)	62m (@ 100km)	500 x 561m (high resolution mode: 500m x 140m) (@ 100km)		
Wavelength	panchromatic 440 - 730nm	5 bands 415, 750, 900, 950, 1000nm	4 bands 1000, 1050, 1250,1550nm	84 bands 512 - 960nm	100 bands 900 - 1700nm	112 bands 1700 - 2600nm
A/D	10bit	10bit	12bit	16bit	16bit	16bit
Base to height ratio	0.57 using TC1 and TC2	< 0.2 using 415nm and 900nm bands	< 0.2 using 1050nm and 1250nm bands	N/A	N/A	N/A

Figure 9.15: Near Future JAXA Lunar Missions

The instruments listed for Selene are very close to the spectral bands used by Clementine, so they will be a very good compliment to Clementine's instrumental data. The spectral profiler is a multi-spectral imager with almost 300 bands and will be operational at the same time as a similar system on SMART-1. The next few years promise to be very exciting for lunar research. It is still to be determined whether or not these instruments will be able to detect nickel/iron M class impact products. It seems that this may indeed be possible, especially in infrared imagers.

Lunar Reconnaissance Orbiter

NASA Goddard Space Flight Center (GSFC) is in the process of identifying scientific requirements for a Lunar Recon Orbiter (LRO) that NASA wants to fly to the Moon by 2008. This is one of the first results of the announcement of the "Vision for Space" by President Bush on January 14th of 2004. There is some information available, even at this early date, concerning potential instruments to be flown on LRO. NASA GSFC has created an "Objectives and Requirements Definition Team," (ORDT), and their charter is to provide NASA's Office of Space Science (OSS) with a *"prioritized set of measurements that can be attained with a resource and schedule constrained LRO mission to be launched before the end of the 2008 calendar year."*[xix]

The ORDT team came up with a list of objectives for the LRO mission, that are presented within the context of the Vision for Space as set forth by the President. These objectives are:

1. Undertake lunar exploration activities to enable sustained human and robotic exploration of Mars and more distant destinations in the solar system.
2. Starting no later than 2008, initiate a series of robotic missions to the Moon to prepare for and support future human exploration activities.
    1. Mission objectives shall include landing site identification and certification on the basis of potential resources.
    2. Measurements shall be made to support applied science and research relevant to the Moon as a step to Mars, engineering safety, and engineering boundary conditions.
3. Technology demonstrations and system testing shall be performed to support development activities for future human lunar and Mars missions.

These objectives are consistent with the new direction for NASA and are a welcome change from the previous lack of interest in the Moon.

To support these objectives, NASA has come up with a set of rankings for the requirements for instruments on the LRO.

    *I: must have for future of human exploration on Moon (**radiation**)*
    *II: must have for landing site selection and safety (**characterization**)*
    *III: must have for **all** next steps in exploration (**geodesy**)*
    *IV: must have for polar resources assessments (**volatiles**)*
    *V: highly desirable to globally assess resources and their accessibility for human exploration (**global resources**)*

NASA as a science-oriented agency is focused on science. There is an interesting set of phrases related to the five requirements above. The first four have the word *"must"* as the operative word while the fifth, which has to do with resources, is framed with the words *"highly desirable."* This is the key issue that has to be dealt with for the return to the Moon. If resources are an afterthought, a lower priority, not equal to the other issues, then NASA will not have the proper emphasis going forward for the economic development of the Moon and its resources. If economic resource considerations are ignored then the value of any NASA science missions are of little long term value to the nation.

Return to the Moon for Economics Sake

This is where the next chapter will be going as the rationale for a return to the Moon is developed. As Bill Clinton said in 1992, "it's the economy stupid," for the return to the Moon, it is the economic impact of resources derived from the Moon that will make the difference between success and failure

Even with the instruments that are being flown on SMART-1, Lunar-A, Selene, and LRO, there is an inadequate amount of data regarding the true potential for resources, especially economically valuable ones, on the Moon. None of these missions are specifically designed to address this issue. None of the architectures that have been proposed for the human return to the Moon have economics as their key focus either. Developing an economically viable architecture is not NASA's goal, nor should it be. NASA is no better at centralized state planning of the space economy than a Soviet Politburo was in the old Soviet Union.

Throughout history, the state has been a macroeconomic enabler. Rome built roads, not only to consolidate their conquests, but also to encourage trade and assimilation into the empire. The British chartered private companies to build canals and railroads, and offered lucrative prizes for technological innovation. The American government strongly supported macroeconomic development by providing the backing for the national railroad, the Panama Canal and the U.S. Interstate Highway system. For a space economy to develop, this level of support will be necessary as well. However, allowing the NASA centers to decide exactly what missions are to be flown, their objectives and the economic benefits, contradicts what the state is capable of doing. Science is wonderful and helps to better all of our lives, but only when the private sector is involved and making the decisions regarding what is economically viable.

*   *   *   *   *

[i] G. Roth, The System of Minor Planets, Translated by A. Helm, D. Van Nostrand Company, Inc., Princeton, NJ, 1962, P. 105-109

[ii] http://www.lpi.usra.edu/expmoon/ranger/ranger.html

[iii] http://nssdc.gsfc.nasa.gov/database/MasterCatalog?sc=1968-027A

[iv] http://www.lpi.usra.edu/expmoon/orbiter/orbiter.html

[v] http://www.lpi.usra.edu/expmoon/Apollo15/A15_Orbital_subsat.html

[vi] http://www.lpi.usra.edu/expmoon/Apollo15/A15_Orbital_laser.html

[vii] http://www.lpi.usra.edu/expmoon/galileo/galileo.html

[viii] Ibid

[ix] http://nssdc.gsfc.nasa.gov/planetary/clementine.html

[x] http://nssdc.gsfc.nasa.gov/database/MasterCatalog?sc=1994-004A&ex=1

[xi] http://nssdc.gsfc.nasa.gov/database/MasterCatalog?sc=1994-004A&ex=2

[xii] E.M. Eliason, et al, A Nea-Infrared (NIR) Global Multispectral Map of the Moon From Clementine, Lunar and Planetary Sciences XXXIV, 2003 (2093.pdf)

[xiii] http://nssdc.gsfc.nasa.gov/database/MasterCatalog?sc=1994-004A&ex=3

[xiv] http://nssdc.gsfc.nasa.gov/database/MasterCatalog?ds=PSPG-00224

[xv] http://nssdc.gsfc.nasa.gov/database/MasterCatalog?sc=1994-004A&ex=4

[xvi] P. Lucy, Lunar Poles, Status of Understanding a Potential Resource, University of Hawaii and Manoa

[xvii] http://www.esa.int/esaSC/120371_index_0_m.html

[xviii] S. K. Dunkin, et al, X-ray Spectroscopy of the Lunar Surface Usign the D-CIXS Instrument on ESA's SMART-1 Mission to the Moon, Lunar and Planetary Science XXXIV, 2003 (1678.pdf)

[xix] J. Garvin, NASA Headquarters Office of Space Science, Objectives/Requirements Definition Team leader, April 2004 (LRO_ORDT_Prelim_FINAL.ppt),

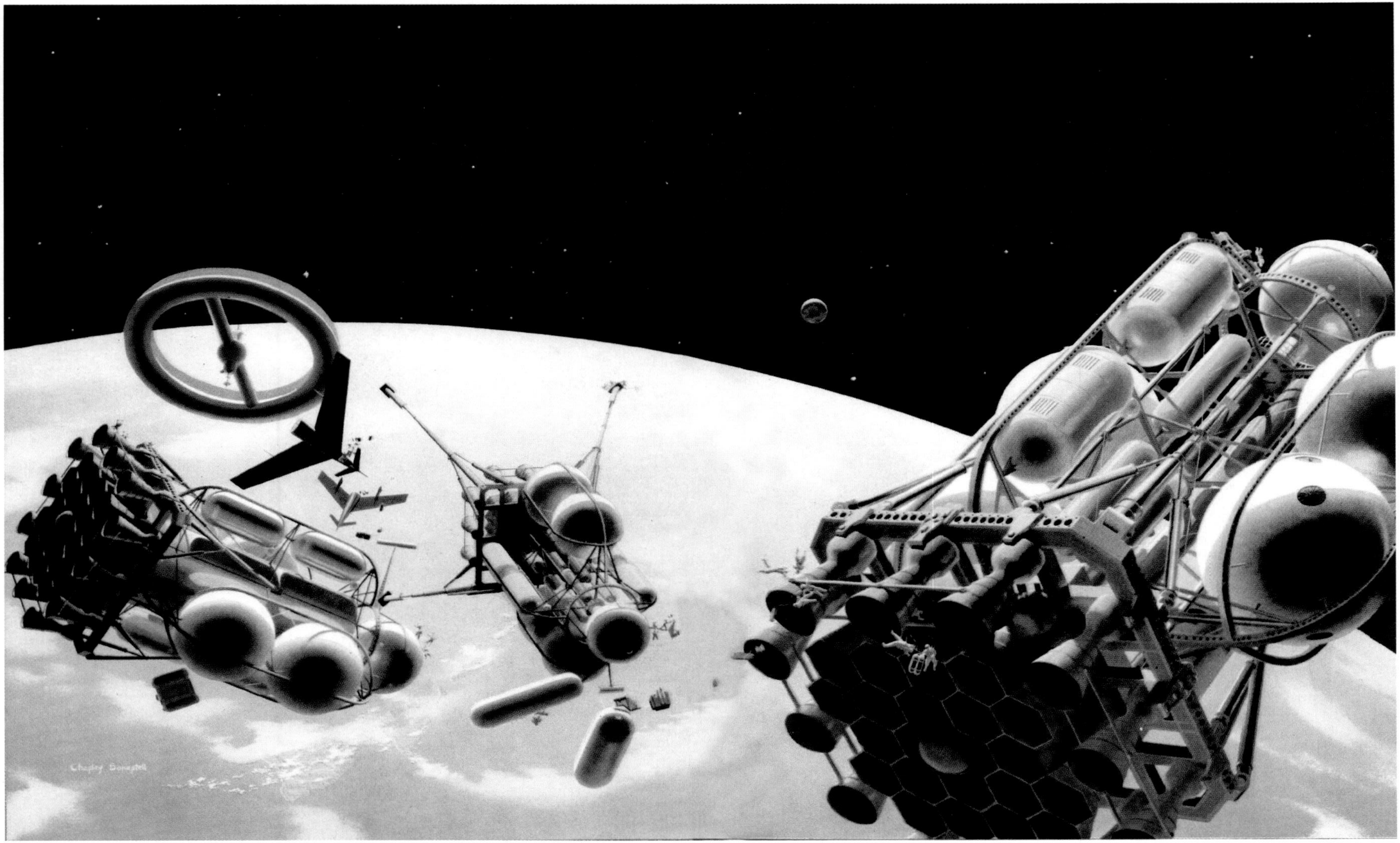

Plate 1. Assembling Ships *(Reproduced Courtesy Bonestell Space Art)*

Plate 2. Lunar Landing (Reproduced Courtesy Bonestell Space Art)

Plate 3. Landers On The Moon *(Reproduced Courtesy Bonestell Space Art)*

Plate 4. Exploration of the Moon *(Reproduced Courtesy Bonestell Space Art)*

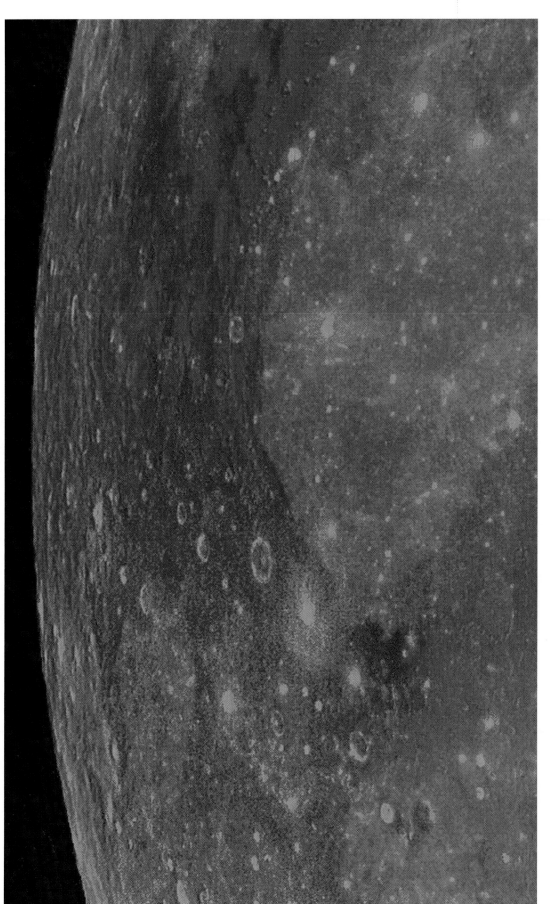

Plate 5. Galileo Flyby Gamma Ray Spectra Map of Lunar Elemental Composition *(Picture Courtesy Lunar and Planetary Institute)*

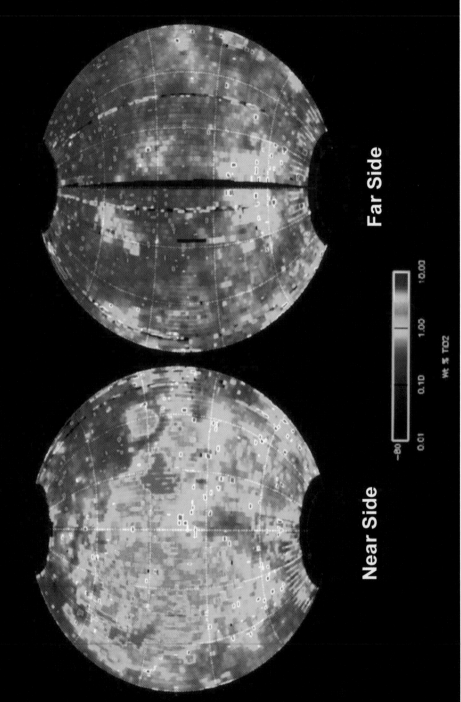

Plate 6. Clementine Map of the Global Distribution of Titanium on the Moon. *(Picture Courtesy Lunar and Planetary Institute)*

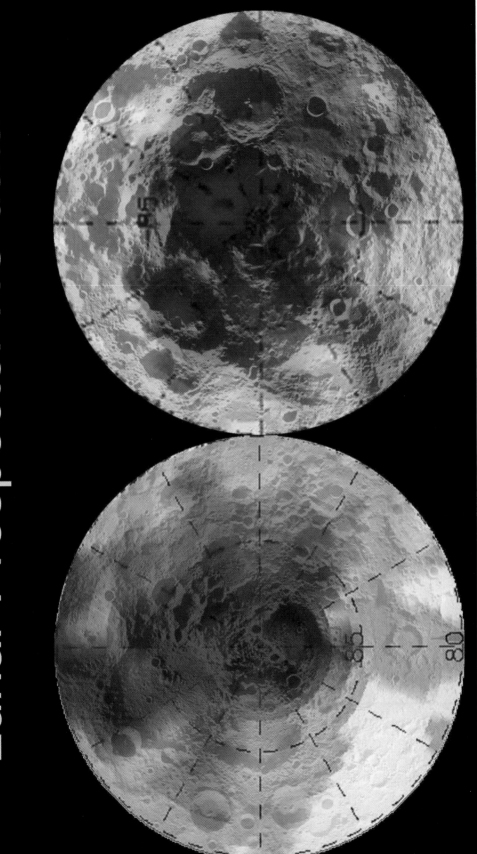

Lunar Prospector NS data

Plate 7. Lunar Prospector Inferred Distribution of Elevated Hydrogen [Water?] *(Picture Courtesy Lunar and Planetary Institute)*

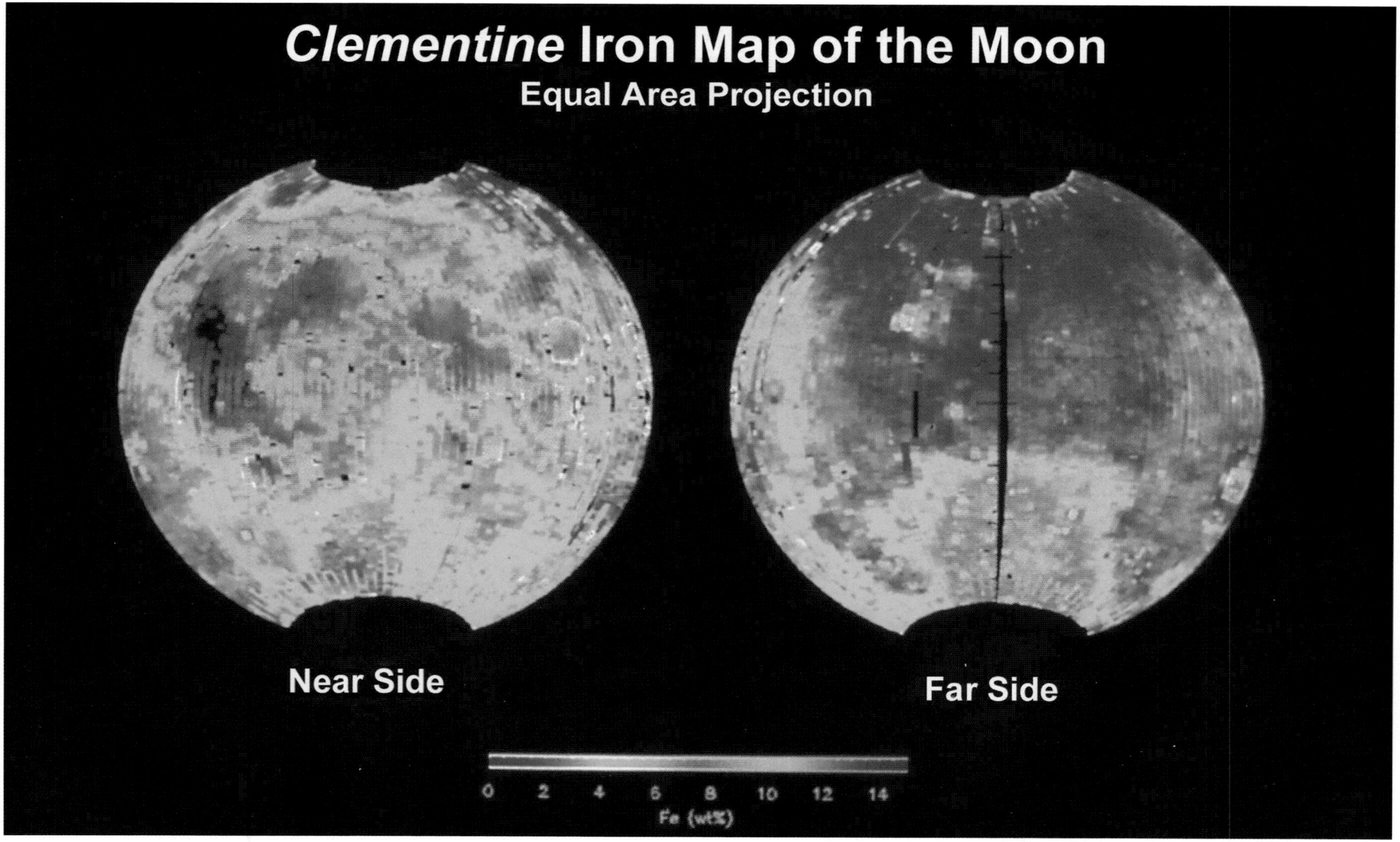

Plate 8. Global Distribution of Iron Derived from Clementine 750 & 950 Nanometer Image Filters *(Picture Courtesy of the Lunar and Planetary Institute)*

Plate 9. Clementine Lunar North Polar Mosaic *(Picture Courtesy Lunar and Planetary Laboratory)*

Plate 10. Clementine Lunar South Polar Mosaic *(Picture Courtesy Lunar and Planetary Laboratory)*

Plate 11. Iron Concentrations from Clementine Data at Apollo 16 Landing Site *(Picture Courtesy Lunar and Planetary Laboratory)*

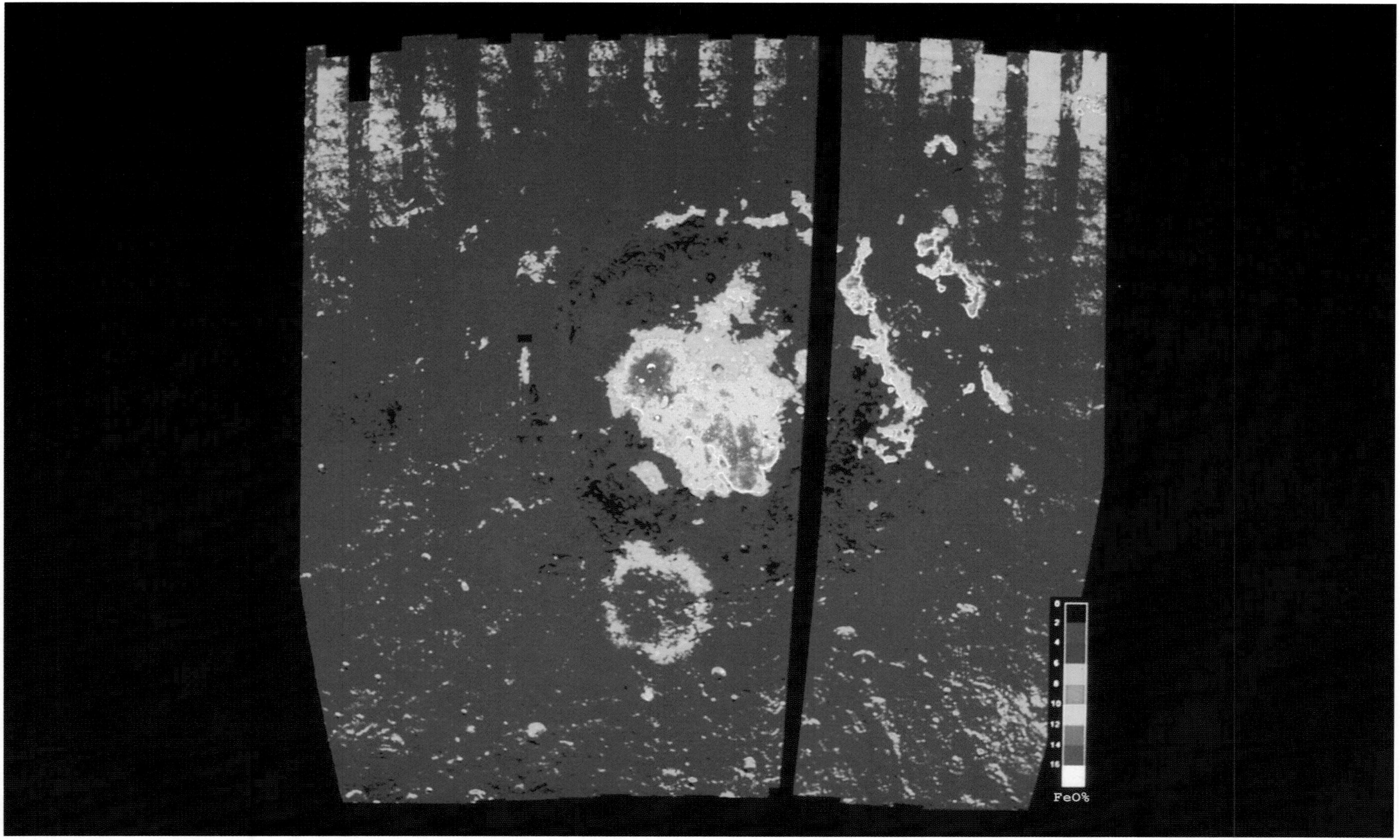

Plate 12. Iron Concentrations in the Oriental Basin from Clementine Data *(Picture Courtesy Lunar and Planetary Laboratory)*

Plate 13. Lunar Lander Being Assembled at the International Space Station (*Picture Courtesy Mark Maxwell/SkyCorp*)

Plate 14. Lunar Landing Vehicle on Final Approach (Picture Courtesy Mark Maxwell/SkyCorp)

Plate 15. Cargo Lunar Lander Being Dissassembled for Spare Parts *(Picture Courtesy Mark Maxwell/SkyCorp)*

Plate 16. Initial Lunar Base for Resource Utilization *(Picture Courtesy Mark Maxwell/SkyCorp)*

Chapter 10:

Lunar Exploration Architectures for the Moon—1953-1972

To develop architectures for the return to the Moon, one must start with a set of objectives. One of the central purposes of this book is to describe how a cost effective overall architecture for the acquisition of material wealth from the Moon and delivered to the Earth for the benefit of the people of the Earth can be constructed. The word "architecture," as defined in the context of this book, means an overall design of the hardware and the method of going to the Moon, landing, and building a lunar economic infrastructure. The major architectures developed in the past have had either exploration or science as the overriding theme. Before I lay out my architectural solution, it is instructive to look at these early architectures and see which of their components can be integrated into a new model that emphasizes economic development. This chapter will look at the history of these architectures and examine why the Apollo architecture ultimately failed the American people.

Historical Lunar Architectures

All endeavors of a purposeful nature involve logistics. Whether planning for a family vacation, the construction of the Great Pyramid, the Panama Canal, or a trip to the Moon, logistical considerations are a critical factor for success. The French in the 1880s were technically capable of building the Panama Canal. The failure to consider the whole problem of logistics, including solving the Malaria problem, doomed their efforts. The Americans succeeded brilliantly by solving the Malaria problem, then used railroads to facilitate the efficient movement of men, machines and material, to enable the Canal to be built in a timely and cost effective manner.

Logistical considerations are crucial when designing architectures for sustainable space development. An example of divergent logistical goals would be the railroads in the United States today. Although railroads were originally developed for passengers and cargo, today cargo dominates rail traffic and highways carry most passengers and short haul cargo. This was due to the logistical efficiency of building and maintaining roads versus rails to go to individual homes and businesses across the nation. The same logistical considerations are pertinent for space logistics architectures. A logistical architecture for a space science based exploration program is significantly different from the logistics of a commercial space development program. The logistics that are adequate for the former fail the latter miserably. This will become evident as we continue.

Lunar exploration and development architectures have been with us for a long time and a lot of thought has gone into them. These architectures have gone through three distinct phases in the past. The first phase is essentially from the end of WWII until the advent of NASA in 1958. The second phase was the Apollo program and its variations, lasting from 1959 through the early 70s. There was a gap in government related interest from the early 70s until the era of the Space Shuttle. There was a lot of very creative work accomplished in this era by public interest groups founded by von Braun and others for the purpose of designing lunar and space settlements. This work was actually in many cases adopted by NASA early in the era of the Space Shuttle up until the advent of the Space Exploration Initiative. Finally the modern era of lunar architecture development commences with the announcement in July 1989 by President George Bush of the Space Exploration Initiative or SEI. We are entering the fourth phase today, one that is still in its formative stage. This chapter will examine the first phase.

The First Age of Space Exploration Architecture

The first age of space exploration architecture development based upon realistic engineering and science began with a series of articles in the popular magazine, Collier's, in 1952. The titles for these were:
Man Will Conquer Space Soon [a collection of eight articles]; March 22, 1952
Man on the Moon, The Journey, and Inside the Moon Ship; October 18, 1952
Man on the Moon, Inside the Lunar Base; October 25, 1952

These articles were by a group of respected rocket scientists including Dr. Wernher von Braun, Willy Ley, Fred Whipple, Rolf Klep, Fred Freeman, Chesley Bonestell and Cornelius Ryan. The ideas presented in these articles were based on hard science and engineering and were the first tightly presented architecture for executing a lunar exploration mission. I call this era, "The Age of Giants," because the ideas in these articles were gargantuan in scope!

Wernher von Braun, the former German, now American rocket scientist, and his peers, designed a three stage fully reusable launch vehicle 265 feet tall and 65 feet in diameter at the base. The 51 first stage engines produced 28 million pounds of thrust (the five Saturn V F1 engines produced 7.65 million pounds). The 34 second stage engines produced 3.5 million pounds of thrust (the 5 Saturn V J2 second stage engines produced 1.25 million pounds). The 4 third stage engines produced 440,000 pounds of thrust (the single J2 engine in the third stage of the Saturn produced 225,000 pounds of thrust). The total weight payload that could be lofted into orbit was 36 tons or 72,000 lbs to an altitude of 1,075 miles. Figure 10.1 is a copy of the original engineering paper drawing of the ship.

The ferry ship was designed to be not only reusable, but extensively so. A fleet of these ferries would first assemble a pinwheel space station in orbit at 1,075 miles. At this altitude, the space station would be out of the Earth's residual atmosphere and would be in a good position to launch large fleets of spacecraft to the Moon and Mars. At the time (this is before the space age actually began in 1957), scientists did not know about the Van Allen radiation belts, so this seemed like a good altitude. We know now that radiation intensity goes up dramatically at altitudes above 800 miles, making life for human occupants unhealthy. This is the reason that the International Space Station today flies at 400 kilometers.

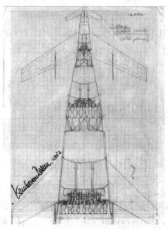

Figure 10.1: Von Braun Ferry Rocket

The pinwheel space station and the spaceship assembly area are shown in plate 1. This station is constructed as a wheel so that it can rotate to provide simulated gravity for the crew. This space station would be used as a scientific research center for microgravity (at the center where the rotation velocity is zero), astronomical research, photo reconnaissance of the Earth over enemy territory (during the cold war) and as a logistics base for assembling fleets of lunar and Mars bound spacecraft. This was a key facet of von Braun's plan because there was no possible way to lift the type of lunar or Mars spacecraft that he wanted to build with a single or even a few launches. This space station would have been 250 feet across, and would have had a crew of 80 men. There were no women in space in the visions of that era.

Consideration of the scale of the lunar spacecraft gives an idea of the impossibility of launching them as one unit. Each lunar exploration lander would weigh 4,370 tons or 8,740,000 lbs. This is more than the Saturn V weighed fully fueled! The lander would measure 160 feet tall (taller than the Statue of Liberty) and 110 feet wide. The tanks are made from fabric and would hold a toxic mix of hydrazine fuel and nitric acid for the oxidizer. The inflatable personnel sphere would be 33 feet in diameter and would have five decks for the 17-man crew's living quarters. Each of the roomy decks would be devoted to a different function.

• *Deck 1:* This is the location of the Bridge, where the captain, pilot, engineer and radio operator run control consoles.

• *Deck 2:* This deck contains the ship's computer, a bulky 50s era behemoth, into which the navigator and his two assistants feed control tapes.

• *Deck 3*: This is the largest deck, containing the living quarters with galley, and sleeping stations.

• *Deck 4: This deck* contains storage and the ship's toilet.

• *Deck 5: This deck* contains engineering, where the life support equipment, batteries, power system components and other critical systems reside. A hatch in deck 5's floor leads into an airlock that opens onto a catwalk 130 feet above the lunar surface.

Chesley Bonestell's original illustration of the lunar ships being assembled in orbit is shown in plate 1. The logistics of the construction of the lunar ships is incredible. Two flights **each day** by the cargo shuttles would bring up 72 tons of components and supplies just for the construction of the three lunar vehicles. At this rate, it would take 80 men in the station 60-70 days to complete each lunar vehicle, or a total of 200 days to build all three. After a test period, the three craft would blast off with their 50 man crews for a six week stay on the lunar surface at Sinus Roris.

The crew of the 50 man expedition to the Moon would be led by a "scientist." Each of the three lunar ships would have a crew compliment of 17. Each of the three ships would have a "captain," one of which would be the commander over all of the vessels. This is the same approach as that of the early 20th century arctic exploration teams. The team would have eight electrical engineers and six mechanical engineers. Since this was in the early days of computers, I guess software engineers were not needed. There was an astronomer, a surveyor and three photographers to record the epic events during the mission. The science team would be composed of a mineralogical team (seven members in various disciplines), a geophysical team (three members) and a team of physicists (three scientists and two technicians).

Prior to sending a manned crew, a cargo vehicle would have been landed on the Moon, carrying three large rovers, two of which would carry 20 men and the other 10 men plus cargo. There were another three pressurized, tracked rovers capable of supporting up to 7 men each for 12 hours at a time. Each of these had a crane that could lift a rover out of danger and could tow up to three trailers. These rovers and trailers are shown in plate 4. A separate cargo lander, based upon the manned design, with a total cargo capacity of 285 tons also carries a surface outpost of prefab Quonset huts. After the three landers set down on the Moon, their cranes off-load additional supplies and then have their empty landing tanks and other hardware no longer needed removed for use at the lunar base.

After the base was constructed, the crew and their rovers would make a trek of 250 miles to the crater Harpalus. During this epic travel across the lunar surface, scientific experiments are performed, data taken and automatic recording devices such as seismometers and other gear would be installed. After all this exploring was over, the crew would take-off with their much lighter lunar landers for the return trip to the Earth orbiting space station.

This is obviously a huge undertaking, but one that the eminent scientists and engineers who wrote the articles in Collier's absolutely thought could be carried out. Logistically, the scale of transportation (72 tons of material a day lifted to orbit) and the size of the lunar vehicles are extremely interesting. In order to completely outfit the three manned vehicles plus the unmanned cargo ship, it would take almost 500 launches from the Earth, not counting the mass of the space station and the logistical support that it would require.

Von Braun understood a key idea that still haunts those today that want to build low cost reusable launch vehicles: the need for high flight rates to amortize the non-recurring costs. If a Boeing 747 only flew six times each year, the ticket price would not be affordable for most travelers. Your car would also be very expensive to own if you only drove it six times a year. This has been the problem with NASA's Space Shuttle. The lack of demand for a high flight rate is a key barrier, even today, holding back the advent of low cost transportation to orbit.

The group that put together these articles also understood another key idea that is also not getting the attention that it deserves today. This second key idea is the importance of a base in orbit around the Earth to simplify logistics and the design of spacecraft destined to land on airless bodies. If we are going back to the Moon for anything other than additional "flags and footprints" missions, the size of manned lunar spacecraft proposed for launch on a single heavy lift launcher is pathetically inadequate to the task. Even the later, mighty Saturn V, with a lift capacity of 260,000 pounds or 130 tons to orbit, could only carry three people to the Moon for only the briefest of times. None of the Apollo lunar missions to the surface lasted more than three days. It was simply logistically impossible to put enough weight in supplies on one launch vehicle for a longer mission. This is the Achilles heel of Apollo and is a logistical bottleneck for any similar proposals to use heavy lift launch vehicles today. The von Braun group's idea of the assembly of the fleet of lunar spacecraft in orbit was the only way to establish the logistical capability to effectively explore the Moon.

It is obvious from the articles and books of the 1950s that the goal of going to the Moon was for its value as an exploration target. The endeavor was organized like a military campaign, staging vehicles and supplies in Earth orbit as well as on the lunar surface. This was necessary for safety and for the unfamiliar environment that the explorers would experience. The Moon was a completely unknown setting then. It was only in the 1960s that scientists finally concluded that the craters on the Moon were from impacts and not from volcanic activity. It is astonishing to one informed of the latest discoveries in our solar system today, to look back at how woefully ignorant we were of the Moon and its resources, as well as the rest of the solar system. This was the real value that the Apollo program brought to the world. It provided an explosion of scientific discovery and technical advance far beyond where we were in the 1950s. When the Collier's articles were written, neither the U.S. nor Russia had launched its first satellite into orbit. We were just as ignorant of conditions around the Earth as we were of the Moon. The discovery of the Van Allen radiation belt was just the first thing we learned that forced a reconsideration of what a lunar exploration architecture would look like. Next, we look at the Apollo era and its architecture for reaching the Moon.

The Apollo Lunar Architecture

When President Kennedy made his historic speech before a joint session of congress on May 25th, 1961, it signaled an immense change in thinking about the "how" of going to the Moon. Up until this point, von Braun and others in the newly formed NASA had looked at going to the Moon in an incremental fashion based upon the knowledge that funding would be limited for such enterprises. After the launch of the Soviet Sputnik in 1957, President Eisenhower increased spending, but was unwilling to be goaded into a massive crash program for space exploration. This is illustrated by a passage in "Managing NASA in the Apollo Era" where it was said:

> *...At the beginning of 1961 the future of NASA's manned space program was uncertain. In December 1960 a PSAC committee had reported that a manned lunar landing was feasible but that it would cost between $26 and $38 billion. President Eisenhower refused to approve any manned program beyond Mercury, save for $29.5 million for a spacecraft for Apollo, as NASA's lunar landing program was designated in the summer of 1960...*[i]

In a January 1961 report to incoming President Kennedy, Jerome Wiesner, an engineer from the Massachusetts Institute of Technology (MIT) and the chair of the President's Ad Hoc Committee on Space, warned the President of "grave deficiencies" associated with the manned space program. The report did not make any recommendations, but set the stage for a fight that would eventually transform NASA and any future space exploration and development architecture developed at the agency. The crucial statement was:

> *The worth of the civilian space program, their argument ran, was proportional to the scientific information obtained.*

The book then goes on to question why the Apollo lunar program unfolded as it did in light of the above argument? In the end, it was the contention that the Apollo lunar landing would demonstrate to the world, in prestige value, the technological superiority of the United States over its Soviet rival. This may seem odd nearly a half century later to those examining the lunar decision, but the Cold War mobilization of the nation's scientific and technical might to achieve the lunar goal was far preferable to a real war with nuclear weapons! In that era of two delicately balanced global powers with thousands of nuclear weapons pointed at each other, the threat of war was a clear and present danger. One that came closer to happening in the early 60s than at any other time in history.

General Bruce Medaris, von Braun's former boss at the Army Ballistic Missile Agency (ABMA), best outlined the philosophical basis for the Apollo architecture:

> *Outstanding feats in outer space are today the greatest advertising medium the world has ever known. If we are to sell our product, our system of freely organized society, to the rest of the world and most particularly to the uncommitted nations, we must advertise our competence and our ability to protect our friends by demonstrated superiority in the penetration of this new and challenging environment. We cannot afford not to compete. We must put our energies into the competition determined to win, and we must admit that we are competing...[ii]*

This is the basic rationale for the lunar program that Kennedy sold to the country in his May 25th speech and reaffirmed on September 12, 1962 at Rice University. With prestige and advertising as the objective and the Moon as the destination, development of the architecture to achieve the stated goals began.

In the early days after the President's decision, a huge amount of work was undertaken by NASA and outside organizations to develop a tenable architecture to get the United States to the Moon by President Kennedy's deadline of the end of the 1960s. There were three architectures that were seriously considered for the trip to the Moon within the context of the fixed deadline. Figure 10.2 illustrates the three methods that were considered:

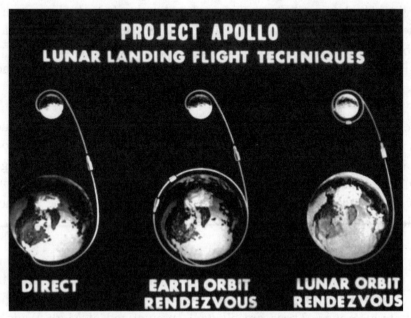

Figure 10.2: Lunar Mission Architectures Studied by NASA (Picture Courtesy NASA)

The three architectures in figure10.2 launched a thousand contracts and an enormous number of man-hours of studies by NASA, industry and academia. To this day, the controversy still simmers about the value of the architecture that was eventually chosen. It is clear though, that, while the chosen architecture was successful in executing an initial reconnaissance of the Moon, it was an utter failure for opening the Moon and beyond to real exploration and development. Examining the Apollo decisions and why they were made can help us learn what not to do for the future.

Direct Flight

The simplest plan was to build a rocket big enough to throw a payload of sufficient size directly to the lunar surface from the Earth's surface. NASA and its predecessor, the ABMA, under von Braun, had been studying different mission architectures for the Moon for quite some time. These studies were the outgrowth of von Braun and Willie Ley's leadership of the team that wrote the articles in Collier's in the early 50s. The fundamental issue was that, in order to get any useful payload to the Moon and back as one single unit, a very large payload capacity was needed, nearly 150,000 pounds to lunar orbit. This would require a huge rocket with over 10 million pounds of thrust at launch. The payload delivered into Earth orbit would be around 350,000 pounds. The project was called Nova. It was on the same scale as the vehicle proposed in the Collier's articles, but not reusable, nor would a fleet be assembled in orbit.

Fortunately, the planners were not starting from a completely clean sheet design for Nova. In 1959, the U.S. Air Force had begun a development program for the F-1 engine, a large liquid oxygen (LOX) and purified kerosene mixture (RP-1) with 1.5 million pounds of thrust. Also, work was under way for the development of a large LOX and liquid hydrogen engine called the J-2. Another LOX hydrogen engine called the M-1 was studied, but never built. Putting these together with different variations of solid boosters and nuclear engines gave the rocket designers their propulsion systems around which to design their vehicles. Figure 10.3 gives a list of the different variations of the Nova for the Moon effort between 1959 and 1962:

The above chart shows the different configurations and compares them to the eventual choice of the Saturn V on the far left. The payload delivered to lunar orbit is the number in kilograms (k) at the bottom of the chart. The numerical designations in figure 10.3 are related to the number of F-1 engines in the first stage. Also, solid rockets were included and are designated by the diameter of the engines and the number flown.

These are by no means the only configurations studied. NASA had a huge number of contracts that were given to all of the large contractors of the time (MacDonnell Douglas, General Dynamics, Martin Marietta), as well as to other independent think tanks and study groups. These are all well summarized at Mark Wade's website, Encyclopedia Astronautica, referenced at the end of this chapter. As shown above, many of the designs lifted less than the Saturn V and would have required Earth Orbit Rendezvous to work. Others required the use of nuclear upper stages, still in the early development phase, to work. Some were basically the world's biggest roman candles proposed by Hercules Powder. The one that was most studied is the 8 F-1 configuration that was called the Nova 8L.[iii] There is an exhaustive history of the Nova and all of its variations at Mark Wade's website.

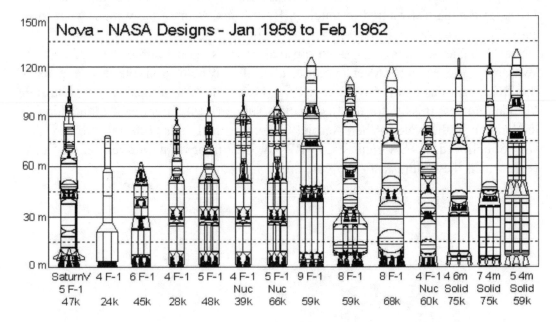

Figure 10.3: Launch Vehicle Configurations Studied by NASA for NOVA

(Illustration courtesy Mark Wade http://www.astronautix.com/lvfam/nova.htm)

Some of the designs by outside contractors were truly astounding systems. Many of these launch vehicles could lift a million pounds to orbit, almost four times what the Saturn eventually launched. One, proposed by Dr. Lyman G. Bonner, director of development at Hercules Explosives Department, a division of United Technologies, was an all-solid rocket motor system with a liftoff thrust of 20 million pounds![iv] To put this into common terms, it was once said that if the Saturn V blew up on the launch pad it would be an explosion half as powerful as the first nuclear weapon. Well if this blew up it would twice as powerful! These would be clustered solid rocket motors with a diameter of 30 feet. This is the most powerful solid rocket motor ever seriously proposed. As a comparison, the Space Shuttle's solid rocket motors that are flying today are 10 feet in diameter.

Another far out yet simple design was from the former naval rocket designer, Captain Robert C. Truax. Bob Truax and his team at Aerojet General did a very detailed study concerning the design of launch vehicles. What they found was that the cost of a launch vehicle had little to do with its size. The cost of the engineering was about the same, the cost of the fixtures to build it did not cost much more with size, and the testing and integration was pretty much the same for small or large launch vehicles. So they designed a minimum cost booster that used one huge rocket engine. This launch vehicle, called the Sea Dragon, was massive. The height was 768 feet, the width was 78 feet and the thrust of the single engine was 80 million pounds or four times that of the Hercules proposal. Figure 10.4 is a slightly smaller version of the Sea Dragon next to the Saturn V.

This big, dumb booster, as it was called, was simple, but it was so big that none of the aerospace companies wanted to touch it. Bob Truax, being a navy man, went to the shipyards where he described what he wanted. They said, "Sure, where is the contract?"[v] The Sea Dragon was so big that Truax's team figured that the costs for developing the launch tower would make the system prohibitively expensive. Therefore, again using his naval experience (he led the team in the Navy that built the Poseidon nuclear missile for the first nuclear submarines), he proposed launching the Sea Dragon in the water. When Aerojet General turned their proposal into NASA, it was not believed. NASA contracted TRW to confirm or refute the results of Truax's work, which they enthusiastically confirmed.

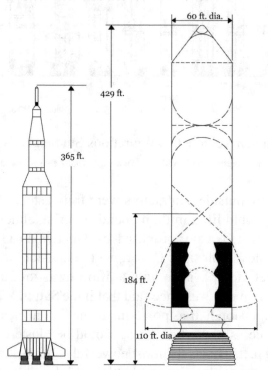

Figure 10.4: Sea Dragon vs. Saturn

During this time of jockeying by different contractors, NASA and the USAF, the design for a direct flight to the Moon seemed to be converging on the Nova 8 variation (the fifth from the right in figure 10.3) and its capability to send 150,000 pounds of payload to the Moon. This would be a five stage design that would lift-off from the Earth and would proceed directly to the Moon using the first three stages without stopping in Earth orbit. The final two stages of the Nova system would land directly on the Moon without any prior circling of the Moon; truly a direct system. The crewpersons would spend their time exploring on the Moon and then would use the fifth stage to take off from the Moon and fly directly back to the Earth with their reentry vehicle.

This architecture was appropriately called the "direct mode," and was championed by the Space Task Group at NASA Langley Research Center in Virginia. This group was to become the core of what became NASA's Manned Spaceflight Center (MSC), later renamed JSC in honor of the late president Johnson. The Nova 8 was an incredible behemoth, with a height of 419 feet (128 meters), 57 feet (17.4 meters) in diameter and weighing almost ten and a half million pounds. This monster had a lift-off thrust of over twelve million pounds. In every way the Nova 8 was an incredible vehicle, but the incredible numbers were its biggest problem.[vi]

The Nova 8 was the least complex logistics solution to the operational problem of flying to the Moon, but it just shifted the complexity to the factory where it would have been extremely complex to design, test and build. In order to build the Nova 8, new factories would have to be built, since there were no buildings anywhere that could handle the size of this vehicle. There were also no test facilities for any of the systems and it would also require the development of a million pound thrust LOX/hydrogen engine at a time when the largest U.S. engine of this type was the 15,000 pound thrust RL10A-1, built by Pratt & Whitney. All of these factors worked together to bash against the hard deadline decreed by Kennedy. Robert Seamans, Associate Administrator of NASA during the early 60s, once told a group of us in a private meeting that Kennedy originally wanted the landing to happen by the end of his second term in 1968 and that Seamans and other NASA officials begged him to remove this from his speech.[vii]

It was eventually decided that the Nova 8 could not be built in time to reach the Moon by 1970, so, on December 21st, 1961, seven months after Kennedy's speech, the Saturn C5, as it was then called, was selected as the preferred launch vehicle to take mankind to the Moon. This decision effectively doomed the direct mode of flight as well as the Nova, the Sea Dragon and every other type of super heavy U.S. launch vehicle, leaving them all to the realm of speculation and regret. Direct ascent was not completely dead because there were several contracts still running based on that architecture and many proponents at various levels of the government. However, by the middle of 1962, all efforts were associated with the Saturn V, and the lunar mission architecture was limited to the capability of that launch vehicle.

Earth Orbit Rendezvous

In parallel with the Space Task Group's work on the direct mode architecture for reaching the Moon, Dr. von Braun and the NASA Marshall Spaceflight Center (MSFC) intensively studied their favored architecture, which was called Earth Orbit Rendezvous (EOR). The study associated with EOR is essentially an updated version of the Collier's article based on propulsion system elements currently being developed, with the additional constraint of landing on the Moon before January 1, 1970. This version of EOR would dispense with the construction of a space station, although, it was still being considered for a later time in the program if the exploration effort continued after the initial landings.

There were several variations of EOR studied between 1959-1961, before the decision to produce the Saturn V was finally made. Many of them used the smaller Saturn I and IB that were under active development before their first flight in 1961. However, these were generally dismissed as part of the argument for the Saturn V. The argument was that the rendezvous and docking of several lunar mission components was too risky and required the development of too much infrastructure in low Earth orbit to meet the number one goal of landing on the Moon by the end of 1969. The reason behind this was that at this time on orbit assembly was only speculation and no one really knew if humans could effectively work to build things in space.

The basic Saturn V oriented EOR plan called for the simultaneous launching of two Saturn V rockets with separate payloads. The first Saturn V would carry a two man Apollo Command/Service Landing Module (CSLM) that would be configured to land on the Moon with the same hardware as proposed for lunar direct mode. The second Saturn V would carry a large fuel tank that would then be used to fuel the empty tanks of the CSLM. After fueling, the CSLM would blast out of Earth orbit to land directly on the Moon, as was the plan for the direct mode.[viii]

While this made the construction of the Nova 8 unnecessary, it added the complexity of rendezvous, docking and fueling in Earth orbit. Von Braun and the MSFC team considered this manageable, but it was not the last word on the subject.

One interesting facet of this discussion was that, even if EOR was used, the final stages for landing on the Moon, and therefore, the configuration of the landing and return system, remained the same as with the direct mode favored by MSC. Several configurations that would fit on top of the Saturn V were studied. Figure 10.5 illustrates several variations of the direct mode/EOR Apollo CLSM, as well as the configuration that finally flew:

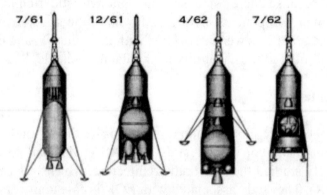

Figure 10.5: Apollo Combined Command-Service/Lunar Excursion Module Designs (Picture Courtesy NASA)

The real reason that von Braun and company wanted EOR was that it provided a path for building a sustainable infrastructure in Earth orbit to support future activities in space. In the words of Dr. Ernst Stuhlinger, a von Braun co-worker:

> *For him (von Braun), the journey to the Moon would not be an end in itself; rather, it would represent but one chapter in the exciting story of man's spacefaring adventures. Unmanned satellites, manned satellites, Earth-to-orbit shuttles, voyages to the Moon, voyages to Mars—it was the beginning of man's reach into space, the transition from homo terrestris to homo spatialis. Operations in Earth orbit, such as rendezvous maneuvers, transfer of propellants, assembly of space vehicles, and also spacecraft modifications and repair, and particularly rescue operations, should be developed. For von Braun, each spaceflight project was part of an organism that grew and evolved almost like a living system. "Let us not consider the journey to the Moon as a 'Kilroy was here' affair," he would say to his co-workers. "This project will be a milestone in man's spaceflight enterprise. Each future project will directly benefit from it."*

The EOR architecture would serve the lunar project and would help to prove technologies for a Mars mission, a space station, deep space staging missions (à la the Collier's articles), maintenance, repair and construction of scientific platforms and many other applications. As this debate raged, time was passing. More than anything, it was the clock that began to change the thinking of many in NASA, including von Braun, to accept another architecture that, while more complicated, had a greater promise of getting the United States to the Moon by 1970. However, with the change, von Braun's ominous prediction about "Kilroy was here," rings down through the decades to haunt us all today.

Lunar Orbit Rendezvous

During 1961 and early 1962, both Wernher von Braun of MSFC and Robert Gilruth of the Manned (Johnson) Spaceflight Center opposed the idea of Lunar Orbit Rendezvous (LOR) espoused by John Houbolt and his team at the NASA Langley Spaceflight center. Each did this for different reasons. Gilruth's team at JSC still lobbied for the direct mode architecture because they felt that this was the lowest risk approach for crew safety. At MSFC, von Braun and team were still solidly behind EOR as the preferred approach. While the December 1961 decision to begin the full-scale development and production of the Saturn V seemed to put an end to the direct approach, this did not keep MSC from promoting it. It is ironic that von Braun opted for an approach that eliminated the super booster and maybe MSC's approach. It is clear that MSFC supported the EOR approach in order to build a longer lasting infrastructure to support a spacefaring nation.

LOR as an architecture had several things going for it in the minds of its proponents.

It did not require that the entire Earth return spacecraft (the one that would launch from the lunar surface and return back to the Earth) be taken down to the lunar surface and later launched back on a trajectory to the Earth. This saved a lot of fuel that would have been needed to lift the extra mass of a single system off the Moon. Taking less fuel saved so much weight that it meant that only one Saturn V would be required. This simplified logistics in Earth orbit, eliminating the need for rendezvous, docking and fueling the lunar vehicle. This in-flight refueling was considered very risky and has still not been adopted by the American space program as of 2004. Von Braun inadvertently aided his opponents by insisting, during the debate on the Saturn vehicle in 1961, that the fifth F-1 engine be included. With the fifth engine, the Saturn had a comparable payload to the Moon as many of the Nova designs. This especially became the case when the F-1 development proceeded very successfully and it was proven to work at even higher thrust levels than its rated 1.5 million pounds (the F-1A version was qualified to 1.75 million pounds). The higher thrust of the existing engines was the equivalent of adding a sixth engine!

With these benefits, both von Braun and JSC's lead systems engineer, the equally legendary Maxine Faget, both savaged Houbolt's plan. According to von Braun, in 1970, Max Faget said that, *"The discrepancy between [this] proposal and the real world is best shown by two figures: Houbolt's fully loaded and fueled lunar excursion module was to weigh a little less than 10,000 pounds. When the spacecraft was actually built a few years later for Apollo 11 that made the first landing on the Moon, it weighed over 30,000 pounds!"*[ix]

According to Ernst Stuhlinger, many at NASA headquarters, as well as President Kennedy's science advisor, supported the EOR architecture. This controversy, downplayed by many today, raged for over a year after Kennedy's call for an American voyage to the Moon. According to Robert Seamans, no other question in the entire space program had more attention lavished on it. According to Seamans, over a million man-hours were put into the question.[x] During 1961, various committees set up to study the issue decided in favor of any one of the three solutions, depending on the biases of those doing the work. This is the beginning of the tension between MSC (now JSC) and MSFC that has persisted for decades to the detriment of NASA and the nation's space program. Actually JSC was still lobbying for the direct mode while MSFC was pushing EOR.

In the LOR mode, the trek begins with the launch of a Saturn V initially into a parking orbit about a hundred miles high. After an orbit or so and after things are successfully checked out, the third stage of the Saturn V would be restarted to accelerate to Trans Lunar Injection (TLI) velocity. After a three day trip the astronauts in the Command Service Module (CSM), with their Lunar Excursion Module (LEM) attached, would fire the engines of the CSM to achieve a low lunar orbit of approximately 60 miles. Figure 10.6 shows the lunar orbit rendezvous according to Houbolt's report:[xi]

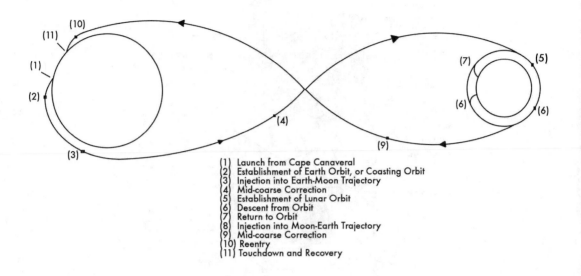

(1) Launch from Cape Canaveral
(2) Establishment of Earth Orbit, or Coasting Orbit
(3) Injection into Earth-Moon Trajectory
(4) Mid-coarse Correction
(5) Establishment of Lunar Orbit
(6) Descent from Orbit
(7) Return to Orbit
(8) Injection into Moon-Earth Trajectory
(9) Mid-coarse Correction
(10) Reentry
(11) Touchdown and Recovery

Figure 10.6: Lunar Orbit Rendezvous Details (Picture Courtesy NASA)

After achieving lunar orbit and checkout of the LEM's systems, the 30,000 lb LEM (Houbolt was wrong by a factor of three as Faget and von Braun recognized) detaches and executes a powered descent to the surface of the Moon, leaving the 45,000 lb CSM in orbit. After spending time on the lunar surface, a period of one to a few days, the LEM fires its upper stage, using the descent stage as a ready made launch platform. A docking in lunar orbit completes the LOR portion as the crew transfers back to the CSM. The LEM is then discarded in lunar orbit, saving a tremendous amount of weight relative to the EOR architecture for the returning CSM. It would have been impossible to send the final Apollo CSM/LEM to the Moon with a 30,000 lb LEM, instead of the original 10,000 lb LEM designed by Houbolt, using the Saturn C4. Von Braun ironically saved LOR and provided for a much better science return for Apollo by his insistence on adding the fifth F-1 engine. This increase in liftoff thrust and improvements in the F-1 engine, allowed for the future inclusion of the Lunar Roving Vehicle and the scientific experiments flown on the CSM in lunar orbit. Figure 10.7 illustrates the Apollo LEM:

At this point in the research for this book, I have come upon a very surprising footnote about the controversy between EOR and LOR. In the original NASA technical document by (NASA TM-74736) Houbolt and the Langley team, it very clearly states that the LOR architecture is equally important and viable even if EOR is chosen as the way to start the mission. The critical passage is reproduced here as figure 10.8:

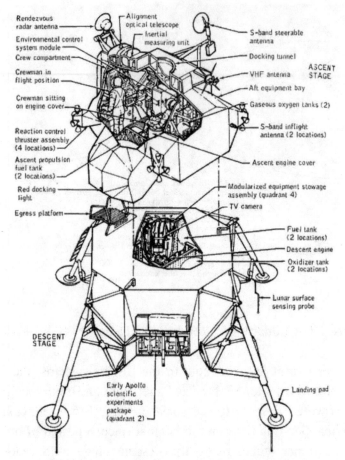

LUNAR MODULE CONFIGURATION FOR INITIAL LUNAR LANDING

Figure 10.7: The Apollo Lunar Module

    2. **Establishment of earth orbit or coasting orbit.-** The lunar orbit
rendezvous mission can be accomplished either with or without earth orbit
rendezvous. In the former case, the benefits of earth orbit rendezvous
can be utilized. In any event, a coasting orbit or coast phase, as
opposed to direct injection, is required to give complete freedom for
injection into the earth-moon transfer trajectory. As discussed in detail
in reference 1, use of a coast phase allows freedom in choice of the time
of the month (i.e., lunar declination) for the mission, allows launch
azimuths within the range safety requirements, allows some freedom in the
design of the trajectory to avoid part of the Van Allen radiation belts,
and permits the design of the trajectory plane to be nearly co-planar
with the moon-earth orbital plane.

Figure 10.8: LOR Use of EOR for Launch?

This passage in figure 10.8 makes it clear that Hubolt was agnostic to the final decision on whether or not to use EOR as long as LOR was used at the Moon.

To someone who has read about this controversy and who knows some of the participants personally, this is a revelation. It is logical that LOR is not dependent upon the Saturn V since it is merely a different method of landing on the Moon, but nowhere that I can find is this brought out in the debates between the various factions regarding EOR vs. LOR. Even more interestingly, in his report, Houbolt remarks on the successful launch of the Saturn I and uses this as an argument for what I call a mixed EOR/LOR mode! While this is a note of historical interest, this seems not to have had an impact (or was ignored) in the debate as MSC swung over to support the LOR/Saturn V and the Saturn V was approved as the launch vehicle of choice for going to the Moon.

The key to this swing by MSC and Bob Gilruth was linked to IBM's progress in miniaturizing the computer system that would be used by the CSM and LEM for navigation. This did not sway von Braun initially. A major objection by von Braun and others to LOR was that many complicated maneuvers would have to take place on the far side of the Moon, out of contact with mission control. These maneuvers included lunar orbit insertion, the initial deorbit burn by the LEM for landing and the trans-Earth injection burn by the CSM at the end of the mission. Another major objection was that the CSM and LEM would have to rendezvous twice, once during the coast from the Earth and once after the landing mission was completed. Yet another objection was one of reliability. There are well over a million parts in the whole Saturn V system. Even if the reliability was a nearly perfect 99.9 percent, an unrealistic number, there would still be 1,000 part failures during a mission. Using only one spacecraft would cut the parts count by a fair margin but add a risky lunar surface separation system. I was told by an old gentleman who was a test engineer on the LEM (and is now a night manager of a casino hotel in Reno, Nevada), that, just to do the separation between the descent stage and the ascent stage at lunar liftoff, all 121 pyrotechnic cutters had to work perfectly or the crew would have died.

At NASA headquarters, the vote was still for EOR. It was not until January of 1962 that von Braun came around and enthusiastically endorsed LOR. This, by no means, ended the controversy. The argument that von Braun used with his own stunned staff was simple. According to Stuhlinger:

> *Von Braun, keeping Kennedy's words "before this decade is out" in mind, became restless. "We cannot afford to continue in this state of limbo. While we talk and talk, the decade is ticking off. Had we been permitted to develop the EOR mode after the president had given the green light for the Moon project, we could have made it by 1969. It is too late now. We better forget, for the time being, the rescue capability in Earth orbit, and the development of a Saturn rendezvous and Earth orbit staging technology. Let's rally our forces behind John Huobolt's LOR mode, and let us put our best and total effort into this project. Bob Gilruth is already backing the LOR mode... The fact that neither MSC or MSFC did originally propose or even support the LOR mode is an extra bonus in this decision. It was Langley's, and in fact John Houbolt's idea and crusade. There need not be any priority complex nor 'not invented here' syndrome either at MSC or MSFC. Believe me, accepting the LOR mode will present us with the only chance to reach the Moon in this decade!"[xii]*

In reading the above, one has to wonder today whether von Braun ever actually read the whole Houbolt report or just went with what he was told of it by the people at MSC. Was this true at headquarters as well? The point of this digression is that NASA's inter-center rivalry was alive and well at that time, which continues to this day to harm the nation's space program. The mixed EOR/LOR mode would have worked and would have enabled what von Braun wanted as well as the benefits that Huobolt's team lobbied so hard to obtain. Von Braun's bosses at NASA headquarters were still very much in favor of EOR. This was also the position of the President's science advisor, Dr. Jerome Weisner. The debate continued to rage for several months until November 7, 1962 when the NASA Administrator, James Webb, gave final approval for the LOR architecture using a single Saturn V.

As we all know, LOR was used and we did successfully reach the Moon on July 20, 1969, five and a half months before the deadline. A second crew on Apollo 12 also made it to the Moon by the end of 1969. We also landed four further crews, with the final crew of Apollo 17 landing on the Moon on December 11, 1972. A footnote of interest is that, if the original EOR mode using a single spacecraft had been selected, the crew of Apollo 13 probably would not have been able to save themselves by using the LEM as a lifeboat and propulsion system. So it seems that an explosion on board would probably have resulted in their death. Thus ended the first chapter of mankind's exploration of the solar system. We are still waiting to return.

Apollo Epilogue

During the debate about EOR vs. LOR mode, von Braun made what is, with 42 years of historical hindsight, a prophetic statement. His quote, *"Let's not consider the journey to the Moon as a **'Kilroy was here'** affair,"* turned out to be exactly the case. Kilroy was the name of a phantom U.S. soldier in WWII who left his trademark, "Kilroy was here," as a slogan across Western Europe and the Pacific Theater. He was a super soldier, always the first to any combat zone, leaving nothing but his name and a drawing of a face peering over a wall. We, who are space advocates today, call the Apollo program, "flags and footprints." In the end, that is the legacy of the Apollo program to our generation who were children when Neil and Buzz walked on the Moon for the first time. Our generation was promised more and we feel betrayed by a nation too absorbed in Earthly matters to see the big picture of the importance of mankind's first landing on the Moon. In a hundred thousand years, if our species survives, and the memory of Vietnam and our other troubles of that time fade to oblivion, mankind will still remember the first time we set foot on the Moon.

As a program, Apollo was as spectacularly successful as the politicians of the time intended it to be: the world's most expensive advertising campaign. They wanted a program to increase our prestige and impress the world. Well…the Apollo program did that, but the Vietnam War abroad, and racial troubles at home, more than blunted this success. The nation turned away from the Moon after the initial landings, so, now, three completed Saturn V launch vehicles and their lunar landers sit slowly rusting away in Florida, Texas, and at the Smithsonian Air and Space museum in Washington DC.

A mantra of the 70s was, "If we can put a man on the Moon, why can't we do X, Y, or Z." What they did not understand in their lack of foresight then was that the exploration and development of the Moon is the answer to solving many of the structural problems that beset our civilization today in 2004. This is also illustrated by the opposition to nuclear power in the 70s that has resulted in more CO_2 pumped into the atmosphere leading to potential climate change. Another loss has been in the disruption in the progress toward nuclear fusion, which is key to our energy future. The loss of the Apollo program cost our civilization three decades in the development and utilization of lunar and other extraterrestrial resources.

The Apollo program did produce some good scientific and economic results. The 384 kilograms of rocks brought back by the Apollo missions still sit in Houston, guarded by NASA, much as the pharaoh's tombs must have been guarded in the first years after they died. We have learned critical things about the make up of the general lunar surface and its geology. We know what the common and some of the rare minerals are in the regolith. We know about some of the resources, such as areas that are highly enriched with titanium and iron. We know how much silicon, oxygen and aluminum is available on the surface. The Apollo samples have provided very important verification for the data gathered by the remote sensing missions mentioned previously. Without the Apollo samples, we would certainly need to get similar samples to determine what possible resources and processes are needed to extract the resources. However, as good as this sounds, it is no better than landing at six places on the earth, none of them near mountains with significant resources, and declaring that we know everything about the geology of the Earth. In the economic area, many of the scientists and engineers who were laid off at the end of the Apollo program, formed the intellectual core that brought us the computer revolution in the late 70s.

The prophetic words of von Braun were fulfilled. Left without any infrastructure in orbit, the Apollo program was easy to shut down. The last gasps of the Apollo program were the Apollo Applications Program that launched Skylab I (Skylab II rests alongside the lunar module in the Smithsonian), only to see it fall from the sky six years later due to atmospheric friction. Three Apollo CSMs were sent to Skylab I, to do experiments and observe the sun. After Skylab, one final Saturn IB roared off the launch pad to rendezvous with our defeated (in the race to the Moon) Soviet adversary as a gesture in a temporary thawing of the cold war. The Soviets, abandoning their aspirations for a manned lunar landing after the explosion of the N1 vehicle, a Nova 8 class launch vehicle, went on to build a succession of space stations. A descendant of these stations today forms the core of the International Space Station, as well as a destination for a few rich space tourists and a few science experiments.

After the Apollo program, NASA offered the soon to be indicted Spiro Agnew three different plans for exploration. All but one involved expanded lunar exploration and an eventual (1986) landing on Mars. This was to be powered by an up-rated Saturn V with the NERVA nuclear engine that was originally to have flown on the cancelled Apollo 20. Billions of dollars were spent designing and building a reusable nuclear rocket engine for extended lunar missions and the flight to Mars, but was never flown. It was shown in a previous chapter. It sits rusting at the U.S. Space and Rocket Center in Huntsville Alabama. All of the ideas studied by NASA for developing a lunar base and going to Mars were shelved and all that was left was the most basic program to build a reusable Space Shuttle. This is the same shuttle that has just cost the lives of another seven astronauts, more than 30 years after its approval by President Nixon and congress, and after achieving very few of its objectives due to chronic under-funding in the design phase.

NASA became the province of the scientific community sometime after Apollo and remains so today. Everything has to be justified for its contribution to science. Profit is not understood or welcomed. Most of the exploration architectures proposed since the death of Apollo have been "science based." None of them have been funded.

The point of this, and the point of this chapter, is that we cannot build a sustainable spacefaring civilization on prestige, other intangibles, or science alone. However, this is what NASA continued to attempt for the next 32 years until this year and President Bush's speech. In the next chapter, I will outline the direction that NASA and a new entrant on the field of exploration architectures, the space advocates. This is the era of the visionaries and the dreamers in the best sense of the word, moving beyond Apollo to the settlement of the Moon and solar system for all mankind. This next generation goes forward until the time of President Bush's Space Exploration Initiative of 1989. The 80s was an exciting era and the one that we will look at next.

*   *   *   *   *

[i] A. Levine, *Managing NASA in the Apollo Era*, National Aeronautics and Space Administration, NASA SP-4102, U.S. Government Printing Office, 1982

[ii] J. Medaris, *Countdown to Decision*, Putnam's Sons, New York, NY, 1960, P. 291-292

[iii] http://www.astronautix.com/lvs/nova8l.htm

[iv] Solid Lunar Rocket Cost Estimates Given, Aviation Week and Space Technology, McGraw Hill, April 2, 1962, P. 18

[v] Personal communications with Bob Truax, Saratoga, California, 1982

[vi] http://www.astronautix.com/lvs/nova8l.htm

[vii] Personal communications with Robert C. Seamans, University of Alabama in Huntsville, 1991

[viii] E. Stuhlinger, F Ordway III, Wernher von Braun, Crusader for Space, Krieger Publishing Company, Malabar FL, 1996, P. 172-173

[ix] Ibid, P. 175

[x] Ibid, P. 176

[xi] NASA-TM-74736, *Manned Lunar-Landing Through Use of Lunar-Orbit Rendezvous*, Volume 1, October, 31, 1961, P. 7

[xii] Wernher von Braun, Crusader for Space, P. 178

Chapter 11:

Lunar Development Architectures for the Moon 1972-1988

The Wilderness and New Beginnings

Studies focused on returning to the Moon after the death of the Apollo program were virtually forbidden at NASA from 1972 until the 1980s. There were a couple of studies toward the end of the Apollo era; one in 1969 based on Apollo hardware and a variation of von Braun's original EOR/Space Station mode, and another one by Rockwell International in 1971 that postulated an architecture based on the capabilities of the Space Shuttle that it was building. However, these architectures became dust and ashes in early 1970s America and the growing indifference to space efforts by the government due to the Vietnam War, Watergate and the first Energy Crisis of 1973-74.

The only organizations in the U.S. that were actively and imaginatively talking about space colonies, the Moon and Mars in the 70s were in the confines of academia and in the emerging space advocacy world. In 1975, Wernher von Braun, singer John Denver, journalist Hugh Downs, Science Fiction writer Ben Bova and many lesser-known space advocates, founded the National Space Institute (NSI), a non-profit educational organization formed to proclaim the value of space to the American people.

The NSI had as its focus, the education of the American people on the merits of a large-scale government space program. In a speech that was read into the congressional record von Braun said:

> ... I know that you are all here ... because you believe, as I do, that a new organization is needed to communicate the benefits of our national space program to the American public. The National Space Institute, which we are formally launching today, shall perform that function. It is a nonprofit, educational, and scientific organization. The main role of the National Space Institute will be that of a catalyst between the space techonologist and the user. It will attempt to bring to the attention of people the new opportunities offered by advances made in space experiments and space techniques. It will study the feasibility of the application, and the potential uses of space technology as it relates to other human activities.

The NSI was basically a marketing agent for the support of a government sponsored space program. As it evolved, they held seminars, conferences, and lobbied congress concerning the value of space for solving earthbound problems. The unfortunate thing is that congress is like the people that it represents, with trends and fads shaping its policy focus. In the mid 1970s when the NSI was founded, space was a fad whose time had past and the government was really not interested in grand efforts in space.

Wernher von Braun died in 1977 and Hugh Downs took the reins of the NSI, but shortly thereafter turned them over to science fiction writer Ben Bova. Bova was an effective communicator through his science fiction books but in the 70s the government was still not listening.

At about the same time as the founding of the NSI, Keith and Carolyn Henson in Tucson, Arizona, incorporated the L-5 Society, based on the work of Gerard K. O'Neill, a Princeton University Physicist. Dr. O'Neill postulated large orbiting habitats filled with tens of thousands of people. Figure 11.1 illustrates his design:

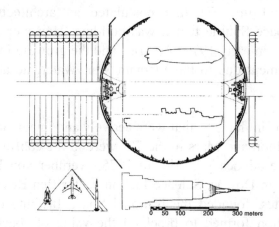

Figure 11.1: O'Neill Colony and Comparative Size of Earth Based Structures
(Pictures Courtesy of the Space Studies Institute www.ssi.org)

In that era of non-support by the government for space activities, the imaginations and the aspirations of space advocates flourished. The founding charter of the L-5 Society (named for a stable location located at the orbital altitude of the Moon) planned a future mass meeting of all chapters at an O'Neill Colony located at the L-5 point, where the organization would then be disbanded having achieved its goal of space colonization.[i]

The L5 Society was much more eclectic in outline and expansionist in vision. The O'Neill colony shown in figure 11.1 was to be the single largest human constructed object in history. All of this started when O'Neill, a physics professor at Princeton University used the construction of such a facility as a thought experiment for his undergraduate students. His calculations whetted his appetite to do more development on the idea, which resulted in a paper on space colonization in 1974.[ii]

We can colonize space, and do so without robbing or harming anyone and without polluting anything.

If work is begun soon, nearly all our industrial activity could be moved away from Earth's fragile biosphere within less than a century from now.

The technical imperatives of this kind of migration of people and industry into space are likely to encourage self-sufficiency, small-scale governmental units, cultural diversity and a high degree of independence.

The ultimate size limit for the human race on the newly available frontier is at least 20,000 times its present value.

The idea that more than just a few government employees living and working on the space frontier was a powerful message in a time when less than 50 people had gone into space. This philosophy is also embedded in this work as we have a more than thirty-year record of the futility of depending on the government to do this for us. Some of the other German rocket scientists were coming to this same conclusion.

The German rocket scientist, Krafft Ehricke, wrote a book in 1978 called, "The Industrialization and Settlement of the Moon." This book was truly visionary in describing the development of a complete industrial civilization on the Moon. Until the end of his life in 1984, he continued to develop his logical thesis, not just for the return to the Moon, but also for the development of an entirely new civilization on the Moon. His idea for lunar development was:

The Moon is the logical proving ground for subsequent industrial developments and settlements elsewhere. Only 2-3 flight days away, it allows us to develop at our very doorstep the experience we need to operate successfully and cost-effectively in more distant regions. No other celestial body and no orbiting space station can more effectively permit developments of the habitats, material extraction, and processing methods, and in essence, all the science, technology, and sociology required for a responsible approach to extraterrestrial operations.[iii]

His grand ideas for lunar development were hardly unique for the time, but his did go farther than most to actually postulate the benefits of a true multi-planet civilization. Unfortunately most space advocates, including the distinguished ones such as von Braun and Ehricke, were (and still are) broke, and while their ideas back then were entertaining, visionary and some even possibly workable, funding from the government or private sources did not materialize during the malaise years of the 70s.

One other group of importance at this time was the one founded by O'Neill himself. The paper that he wrote for *Physics Today* in 1974 spawned further papers and later some small grants from NASA. O'Neill used these grants to further refine the ideas and this work became the basis for his book, "The High Frontier." In the late 70s the financial support from NASA waned and so O'Neill solicited private support, and with two donations totalling $100,000 he founded the Space Studies Institute. Figure 11.2 illustrates what the interior of the Island one space habitat designed by O'Neill would look like.

Figure 11.2: O'Neill's Island One Habitat Interior (Picture Courtesy www.ssi.org)

The Space Studies Institute or SSI was founded as a forum for those who wished to develop the technologies, social policies, and activist means to bring the high frontier about. The Institute achieved some significant technical firsts that were and still are applicable to space development.

SSI held conferences where subjects of interest to space developers such as lunar and asteroid mining, mass drivers, and solar power satellite construction were seriously discussed by people with the technical know-how to pull it off (just no money). Figure 11.3 illustrates some of the activities from the SSI Space Manufacturing conferences.

Figure 11.3: Lunar Mining Installation and Orbital Factory for Solar Power Satellites

On the government front, things began to change after the election of Ronald Reagan in 1980 and the first launch of the Space Shuttle in April of the following year. It does seem that, during the early 80s, there was an impact from the space advocacy movement that began to be reflected in some of NASA's space station studies. NASA had been doing some very low level studies about the "Next Logical Step" after the space shuttle during the late 70s and early 80s. Figure 11.4 shows one version of NASA's idea for logical progression into space:

Figure 11.4: NASA's "Next Logical Step" Progression From the Early 80s

It is striking that the NASA plan as illustrated above does not include a return to the Moon. As was usual, NASA, MSFC and JSC were walking divergent paths into the future with their space station ideas. Most of the work between 1975 and 1984 at JSC was associated with the servicing of GEO platforms and a return to the Moon, while the engineers and scientists at Marshall were focused on Earth observation and microgravity research, much of which was a spin off of the Skylab program.

Activity at NASA increased when President Ronald Reagan, in his 1984 State of the Union address, announced his approval for the development of what later became known as Space Station Freedom. A Low Earth Orbit, or LEO, space station was the central assembly location originally described in the Collier's articles more than 30 years previously. Figure 11.5 provides the pivotal statement by Reagan:

Figure 11.5: Reagan's Call for a Space Station 1984

The decision by Reagan, against the advice of the scientific community and the military, initiated work on a space station that would be all things to all people. It would be a location for the Earth remote sensing and microgravity favored by Marshall. It would also be a base for a space tug for GEO orbit as well as lunar and Mars Missions favored by JSC. However, the struggles were just beginning. For the first time at the presidential level, the term "economic gain" becomes part of the discussion about space exploration and development. For the first time since the death of Apollo, NASA had executive endorsement and at least some money to study space stations and their potential for science, commerce and exploration. Studies, albeit well funded ones, were all that were approved initially as a concession to the program's opponents.

With the advent of the Space Shuttle, the sole surviving part of the 1969 NASA exploration plans, the first part of von Braun's original Collier's article was put into place. Now, with Reagan's stamp of approval for a space station, the second element of that architecture was initiated. All return to the Moon architectures by NASA for the rest of the decade and beyond would follow that path. Several studies looking at space station designs were begun, most of which would support commercial and industrial activities as well as the return to the Moon. Funding also began to flow from NASA to look at architectures for exploration, both for return to the Moon and for manned missions to Mars.

The first truly significant re-examination of lunar base and mission architectures came in the form of a symposium and the resulting set of papers that were truly remarkable. This was a public symposium hosted by the National Academy of Sciences in Washington, D.C., on October 29-31, 1984. This symposium was mostly a showcase for the works of the dreamers and the advocates coupled with the blessings of NASA and the White House. The symposium proceedings shown in Figure 11.6 contained several papers on a wide variety of themes:

Following is the table of contents from the proceedings:

Keynote Speeches
Lunar Base Concepts
Transportation Issues
Lunar Science
Science On The Moon
Lunar Construction
Lunar Materials and Processes
Oxygen: Prelude to Lunar Industrialization
Life Support and Health Maintenance
Societal Issues
Mars
A Vision of Lunar Settlement

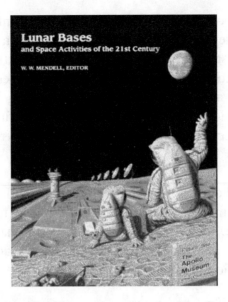

Figure 11.6 Lunar Bases in the 21st Century

This symposium and the ideas embodied in the proceedings listed in Figure 11.2 has become the modern baseline for almost all lunar and Mars related architectures that have development and settlement as their goals. As can be seen from the chapter titles, some exciting ideas for exploration and space settlement were discussed. Talks about lunar industrialization, mass production of oxygen for propellant, lunar construction and lunar settlements, revived the most optimistic dreams of the Apollo era. In the 60s, it was the politicians who pushed the lunar effort as the world's largest advertising campaign. The scientists and the dreamers who wanted more were pushed to the side. After the symposium, it began to seem that the space advocates and NASA insiders who wanted a spacefaring civilization were being encouraged by the government to lay the groundwork for achieving the dream.

This was supported at the highest levels of government as evidenced by the keynote address for this symposium, given by the NASA administrator, James Beggs. The follow up speaker was George Keyworth II, science advisor to President Reagan, and Director of the Office of Science and Technology Policy (OSTP). NASA Administrator Beggs made reference to a National Academy of Sciences Space Science Board Lunar Base Working Group meeting held at Los Alamos National Laboratory in April 1984 that endorsed a lunar base as NASA's long-term goal for the 21st century. Beggs made the following observation about lunar development:

> *I believe it is highly likely that before the first decade of the next century is out, we will, indeed, return to the Moon. We will do so not only to mine its oxygen-rich rocks and other resources but to establish and outpost for further exploration and expansion of human activities in the solar system, in particular, on Mars and the near-Earth asteroids.*

Beggs went on to talk about the development of a permanently manned space station as the next step in the development of space. The space station would be equipped with a "supporting infrastructure" that would enable operations between Low Earth Orbit (LEO) and Geostationary Earth Orbit (GEO) and eventually to the Moon and beyond.[iv] In closing, the administrator spoke of developing the technologies to mine the Moon, reaping the "rich dividends and enormous benefits" that this development would bring. This is very much in keeping with what Krafft Ehricke said in his remarks quoted earlier.

This was followed by the keynote of George Keyworth II, the President's Science Advisor, who sounded a more general theme of exploration for national "inspiration", practical value on Earth, and for education. Keyworth challenged the audience to think about what would come after setting up a lunar base and to think about how to sustain the vision in the long term.

1984 Architecture A

With the Space Shuttle to move crew and high value cargo, and a revived Saturn V or Shuttle derived Heavy Lift Vehicle (HLV) planned, the space station would be the staging and assembly point, just as in the Collier's scenario, albeit with far fewer launches. Figure 11.7 lays out a mission design based on departure and return to the Space Station Freedom:

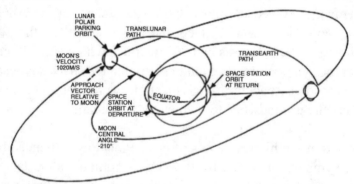

Figure 11.7: Space Station Based Mixed EOR/LOR Lunar Mission Architecture

The scenario as described in a symposium paper by Gordon Woodcock, at that time, a senior scientist at Boeing, is basically a modified version of John Houbolt's original 1961 proposal, but with modern hardware and a space station. Instead of a direct flight by a single Saturn V, several assembly flights would be integrated at Space Station Freedom using multiple Saturn class vehicles combined with the Space Shuttle. A reusable space tug would transfer heavy payloads to and from Space Station Freedom to a Low Lunar Orbit (LLO) where a lander would detach from the orbital tug and then proceed to the surface. After mission completion, whether an independent mission, or a mission to an established lunar base, the lander takes off, returns to the space tug still in lunar orbit, which then transits back to the Space Station. [v]

Woodcock went on to describe in detail the energy that is necessary for these transit orbits, building on his vast experience in developing these types of trajectories (I have had the pleasure of knowing Gordon for almost 20 years). The energy budget in meters per second of velocity increments is shown in Table 11.1:

Mission Segment	Rough Velocity Change (dV)	Calculated dV
Trans Lunar Injection	3,200 Meters Per Second (m/s)	3,139 (75 m/s g loss)
Lunar Orbit Insertion	800 m/s	867 m/s
Landing	2,100 m/s	2,100 (from Apollo)
Ascent	2,000 m/s	2,000 (from Apollo)
Trans Earth Injection	800 m/s	906
Earth Orbit Injection	3,200 m/s	3,061 (propulsive)
Aero-Assist Difference	3,200 m/s	200 (Aero-assisted)
Total	*12,100 m/s*	*12,073 (Propulsive)*
Total With Aero-Assist	*9100*	*9,180 (Aero-assisted)*

Table 11.1: Velocity Increments for Circa 1984 Lunar Mission Architecture (Woodcock)

Gordon Woodcock's numbers form a baseline for a space station-to-low lunar orbit architecture for returning to the Moon. A lot of design work by NASA and by space advocates used similar approaches during this period. Woodcock and others did introduce one new feature that had not been a part of most previous efforts. That difference is the use of aerobraking to save on propellant mass. Aerobraking is using the friction between the vehicle and Earth's atmosphere to slow down a spacecraft, as shown in Figure 11.8. In Table 11.1 above, it shows that by using aero-assist, or aerobraking, as it is commonly called today, a lunar mission can save about twenty-five percent of the total fuel that would be needed if aerobraking were ignored. This is a significant improvement in the amount of fuel that has to be carried all the way to the Moon, but it does not come without a price. Using aerobraking requires the vehicle to carry a heat shield.

Another paper in the same proceedings, by Stephen Hoffman and John Niehoff, was concerned with the overall architecture and included the design of a preliminary lunar base beginning with a transportation node in the form of a space station in low Earth orbit. They also added aerobraking. An aerobrake is basically a heat shield that is mounted to the spacecraft that is returning to the Earth. Aerobraking trades friction for energy in slowing down a spacecraft as it returns from the Moon or other orbit to get it into a low Earth orbit.

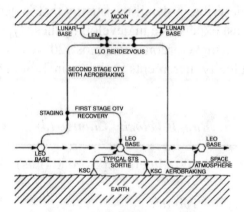

Figure 11.8: Lunar Transit System

In Figure 11.8, Hoffman and Niehoff use a two-stage space tug in a fashion similar to that in Woodcock's architecture. Both the boost tug and the lunar return vehicle use aerobraking to slow down the returning spacecraft so that it can return people or cargo to the space station.[vi] This is different than Apollo Earth or Lunar Orbit Rendezvous that returned the crew directly back to the Earth. The benefit of both of these approaches is the use of the aerobrake. A subtle but important implication of aerobraking involves the weight of the aerobrake. In order to slow down in a single orbit, you have to dip down fairly far into the atmosphere. This means that you have to have a very good (which means heavy) aerobrake. You don't want to carry this extra weight down to the lunar surface, which means that you leave the tug with the aerobrake in low lunar orbit where it waits for the separate lander to return. This does add some complexity to the whole architecture, but the savings in fuel, which weighs a lot more than the aerobrake, offsets this. It is also important not to dip too far into the atmosphere or unfortunate side effects occur!!

There is one very significant difference between what Woodcock was proposing and the architecture of Hoffman and Niehoff. Woodcock has been a proponent over the years of the use of Lunar Oxygen, or LUNOX, as it is commonly called. While others had proposed this, Woodcock quantified the benefits, which are considerable, to an overall lunar architecture. Woodcock and others, who build their architectures around the production of oxygen on the Moon, show considerable savings in total mass lifted from the Earth because 70% of that mass is propellant. If you use lunar derived oxygen, you can save over 50% of the mass lifted from the Earth. This is extremely important at today's launch costs of $10,000 per kilo. This will be discussed further in a later chapter.

Wendell Mendell, Mike Duke and Barney Roberts of the NASA Johnson Spaceflight Center, proposed a very interesting related architecture. Figures 11.9a and 11.9b show their approach:

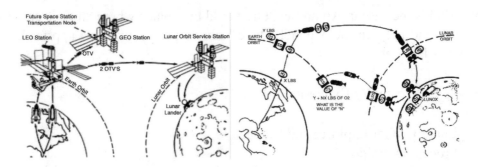

Figure 11.9a: Cislunar Architecture, Figure **11.9b:** LUNOX Added to 11.7a

The overall architecture shown in Figure 11.9a includes a manned GEO space station to service GEO communications satellites, a very interesting foray into commercialization that JSC flirted with in the early 80s. [vii] The difficulty is that NASA was actually ahead of where the commercial industry was then. At that time, the largest communications satellites were the Hughes (now Boeing) 376 series of communications satellites. At the time of the paper (1984), existing launch vehicles were adequate to the task and the advent of very large (>10,000 lb or 4,000 Kg) spacecraft was more than a decade in the future. While NASA in the early 80s launched the bulk of U.S. communications spacecraft and actually serviced a few as well that were stuck in LEO, the idea of on orbit assembly of communications satellites for GEO did not catch on in any meaningful way with the industry. Figure 11.9b illustrates how a space station in Low Lunar Orbit (as opposed to Woodcock's orbiting tug) would be used as part of a very robust and capable architecture incorporating LUNOX.

LUNOX would not only be used to save transporting fuel from Earth to lunar orbit for the return trip, but the depot in low lunar orbit would be a staging point to store and ship this fuel to the GEO station and or all the way back to LEO. This oxygen would be used to fuel tugs for the LEO-to-GEO traffic as well as back and forth to the Moon. This was a very comprehensive and economically interesting architecture for developing a cislunar (incorporating the Earth and the Moon and all space in between) economy. In the end, the Mendell/Duke/Roberts paper had the same goals as Krafft Ehricke: a self sufficient, industrialized lunar economy and society. The same was true of the O'Neill inspired L5 society.

Lunar Bases and Lunar Exploration Architectures 1984

In the "Lunar Bases and Space Activities of the 21st Century" Symposium, NASA, the dreamers and the advocates went far beyond the simple ideas of the 1960s that were focused on the task of getting a few government astronauts to the Moon and returning them safely as part of the world's largest advertising campaign. These visionaries, both inside and outside of the government, wanted to create a "polyglobal," as Ehricke termed it, civilization on the Moon. This would not have just a few government employees as scientific tenants of a government owned facility as was the case in

Apollo, and is the case in Antarctica, but would be a place where citizens could visit, live, work and play. In the Mendell/Duke/Roberts paper, a phased approach to this colony was laid out.

Phase I: Preparatory Exploration

   Lunar Orbiter Explorer and Mapper
   Process Definition
   Site Selection
   Automated Site Preparation

Phase II: Research Outpost

   Minimum base, temporarily occupied, totally resupplied from Earth
   Surface mining pilot operation
   Lunar oxygen production pilot plant
   Closed systems research module

Phase III: Operational Base

   Permanently occupied facility
   Expanding mining facility
   Lunar agriculture research laboratory
   Lunar materials processing pilot plant(s)

Phase IV: Advanced Base

   Lunar ecology research laboratory
   Lunar power station-90% lunar materials-derived
   Agriculture production pilot plant
   Lunar manufacturing facility
   Oxygen production plant
   Lunar volatile extraction pilot plant

Phase V: Self-sufficient Colony

   Full-scale production of exportable oxygen
   Volatile production for agriculture, Moon-orbit transportation
   Close ecological life support system
   Lunar manufacturing facility: tools, containment systems, fabricated assemblies, etc.
   Lunar power station-100% lunar materials-derived
   Expanding population base

This is just a sample of the fertile imaginations of the advocates and dreamers during the early 1980s. Most of these advanced ideas revolve around the following system elements.

Earth Orbit Space Station
NASA's Space Shuttle
NASA Follow on Heavy Lift Vehicle (Shuttle C or equivalent)
Lunar Base
Nuclear Power
Some Variation of Rendezvous and Docking in Low Lunar Orbit
Aerobraking to Improve System Performance
Lunar Resources Utilization

All of these ideas were predicated on the one thing that eventually meant their downfall: government funding. The funding required to build these colonies of hopes and dreams never materialized. Some parts were funded. We have a Space Shuttle that has twice spread its components and personnel over wide areas. This Shuttle was so under-funded at the beginning of its development that any hope of it achieving its goals of low cost, rapid turnaround were killed, leaving a pale shadow upon which to base our dreams. We also have a space station of a sort. One that, while it is actually close to some of the better designs, is in danger of falling apart due to the destruction of the Shuttle orbiters that it so depends on for continued construction and maintenance. This process did not happen in a day, but seems to be, at least partially, the result of the internal struggles of the larger space community, NASA and the advocates, as well as external forces.

Science vs. Development

Scientists have never been all that thrilled about going to the Moon. In 1965, Neil Ruzic wrote a book called, "Case for Going to the Moon." In this book, he developed a survey that queried several thousand scientists and industrial companies regarding their opinions about the value of the Moon to science and industry. Surprisingly, 85% of the scientific community who responded to the survey were opposed to spending money to land on the Moon for scientific purposes. The overwhelming opinion was that they could make better progress for science overall by funding *their* research.[viii] Interestingly, the industrial interests were considerably more positive in their replies.

As the Apollo program wore on, the scientific community did get more involved, especially in the geophysical community. The Apollo era even gave birth to a new cadre of scientists called, "planetary scientists," plus other fields related to studying the physics of the Sun, the solar terrestrial magnetic fields, and other space related research. Ironically, this did not help the scientific rational for the Moon since most of these new scientific fields depended on unmanned probes for their data and the Moon was considered by NASA to have been explored. This just created a new division in the scientific community as the planetary scientists saw (and many still do) manned spaceflight as taking money from their projects to spend on projects with (in their minds) little scientific value.

In early 1984, a National Academy of Sciences Lunar Base Working Group got together in Albuquerque to look at the potential for science in the return to the Moon. At this gathering, they listed potentials for research and also voiced concerns about development. Here is a list of their science topics related to the Moon:

Thermal, Magmatic and Tectonic Evolution
Bombardment History
Evolution of Volatiles
Magnetic Field History and evolution
Analysis of Regolith for Long Term Studies of the Output of the Sun
Scientific Observatories
Solar
Radio
Large Aperture Optical

The scientific community was very interested in the possibility of constructing large observatories on the Moon that would be used to spend large amounts of time staring at objects such as the Sun. Also, a radio telescope on the lunar far side was a high priority for them because its location would block radio signals from the Earth. Constructing telescopes on the Moon would allow for much greater sensitivity than would otherwise be possible on the Earth.

The action of winds, waves and volcanism has erased almost the entire early surface of the Earth. The Moon preserves the early history of the bombardment of the Earth soon after its formation. Due to the obvious difference in cratering rates between the lunar highlands and the Mare, it has been postulated that an "early bombardment" period happened a few hundred million years after the formation of the Earth and the Moon. Even the formation of the Moon and Earth, theorized to be the result of the collision of a Mars sized body with the early Earth, could be further verified by the intense study in deep craters of the composition of the Moon. The Lunar Base Working Group's report was geared exclusively to scientific studies, but they did address development, although this was in a negative light.

It was stated that lunar development activities would seriously compromise the scientific goals. This is a quote from the report.

Most lunar scientific activities require that the unique lunar environment be preserved. Lunar base operations might affect this environment in adverse ways, especially if industrial operations expand.[ix]

From the website where the above was quoted:

Specific potential environmental impacts were cited: increased atmospheric pressure, which could change atmospheric composition and compromise astronomical observations, and increased very low radio frequency background through satellite communication networks, which could affect the use of the far side of the Moon for radio telescopes.

Unprotected by any atmosphere, the Moon will accumulate scars of impacts by humans at an increasing rate...

Extensive mining efforts on the Moon, however, could scar its surface irreversibly.

Richard Tangum expanded upon these environmental concerns as he wrote about the impact of lunar resource utilization for oxygen production. He estimated that it would take the mining of 100,000 tons a year of regolith just to support the generation of 1,000 tons of oxygen. This would take about 50,000 cubic meters of regolith per 1,000 tons of oxygen produced. He went further to make the assertion that a lunar civilization as postulated by Krafft Ehricke would need to strip mine a swath of the lunar surface seven kilometers square and five meters deep. As a further derogatory illustration of this, a drawing of a strip mined Moon was presented as the ultimate fate of the Moon as a result of unbridled development. A similar picture is shown in figure 11.10:

Figure 11.10: The Moon Disfigured by Development

(Picture Courtesy http://pdphoto.org, as Modified by Nikki)

First of all, the numbers quoted for the amount of regolith needed to generate oxygen are completely incorrect. The Moon is almost 40% oxygen by weight and there are several methods of liberating enough of that oxygen to satisfy any level of civilization that we would desire. Second, this concept, as illustrated above, casts those who wish lunar development as despoilers of a natural world.

The implication of this thinking is that lunar development is something to be halted so that the Moon can be preserved for science. While this is a noble sentiment, the scale of development indicated would be hundreds of years in the future if at all. The Moon is an interim source of resources until such time as our civilization has advanced enough to utilize the resources of the asteroids, which are a million times larger and more accessible than those on the Moon. The Moon is more valuable in the near term because it is closer to us in space, which is important to human safety as we learn how to live and operate in this new environment.

More recently (1997), The National Academy of Sciences (NAS) Commission on Physical Sciences, Mathematics and Applications (CPSMA) Space Studies Board (SSB) Committee on Human Exploration (CHEX) wrote a very thorough report that expanded upon the earlier work of the NAS Lunar Base Working Group. Some excerpts from that report give hints about their interests and biases:

*Is science then **the** motivation for a Moon/Mars program? This question was answered in the negative by the National Academy of Sciences and the National Academy of Engineering in a report on space policy prepared in 1988. It stated that "the ultimate decision to undertake voyages of human exploration and to begin the process of expanding human activities into the solar system must be based on nontechnical factors."[x]*

It is interesting that most of the interest generated in the report was from the life sciences community and research related to the adaptation of the human body to the new environments of the Moon, Mars and free space. The report goes on to say the following about lunar and Mars science in general:

Participation of scientists in a program of human exploration is a sensitive subject in the broad scientific community. Some individuals fear that any involvement is an implicit endorsement of such a program. Others fear that science is or will be used as a justification or that low-priority and/or low-quality sciences will be funded under the umbrella of an expensive human spaceflight program...[xi]

The above attitude should not be surprising considering the overwhelming disinterest of the broad scientific community toward the Apollo program of the 1960s and 70s. Here are more of their concerns:

CHEX reiterates the earlier position of the Space Studies Board that a program of solar system exploration that includes only the Moon and Mars and their immediate vicinity as scientifically incomplete. The obvious concern is that a program of human exploration, which by its very nature would be expensive, could dominate NASA budgetarily, managerially, and programmatically to the detriment of a balanced scientific program.

The above continues the humans versus robots argument that has divided the space community for decades now. The implication is that just as much money should be spent on robotic exploration as human exploration. This, while desirable, is not possible within the budgetary realities that NASA and space exploration have labored under. However, there have been several studies that show that, without a manned space program to give context to the robotic exploration, space science would be funded at a fraction of its historical and current budget. This turns the quote from the movie, "The Right Stuff," around. While during the race to the Moon, it was "no bucks, no Buck Rogers," it is more appropriately over time to say, "no Buck Rogers, no bucks!"

The CHEX report goes on to laud the value of observatories on the Moon and basically the same science as described previously, but they did not address any issues related to development. Actually, this is proper. This was a report from a prestigious scientific committee addressing scientific concerns. However, this is also the problem. In the 1980s, scientific concerns and justifications increasingly dominated NASA's thought patterns. It did not happen immediately, but it happened decisively as time moved forward.

Lunar Bases and Exploration Architectures 1986

The National Commission on Space

The National Commission on Space report in 1986 represents the maximum expression of the hopes of the dreamers and space advocates. The cover of the report, shown here in Figure 11.11, illustrates the outcome that the dreamers, represented by former NASA Administrator Tom Paine, Gerard K. O'Neill, Moon walker Neil Armstrong, Dr. David Webb and the legendary pilot Chuck Yeager envisioned. The politicians were represented by UN Ambassador Jeane Kirkpatrick, General Bernard Schriever, as well as congressional advisors, including former astronaut, John Glenn. Dr. Louis Alvarez was the leading scientific mind, who had recently postulated an asteroid impact as the cause for the demise of the dinosaurs.

Figure 11.11: The National Commission on Space Report[xii]

Reading the National Commission on Space Report (otherwise known as the Paine Report) today, one is struck by its incredibly expansive and visionary tone. The report's beautiful graphics and compelling ideas gave the advocates and dreamers in the larger space advocate community much hope for the future. Figure 11.12 illustrates the design of Space Station Freedom as a base for satellite servicing and a locale for the assembly of GEO and lunar bound spacecraft:

Figure 11.12: Expanded Dual Keel Space Station (picture courtesy Marcus Lindroos)

The Space Station Freedom would be the linchpin of an ambitious plan to return to the Moon, go on to Mars and send humans to a selection of Near Earth Asteroids. The plan expanded on previous work and advocated the use of lunar oxygen and the iron from meteorites known to be bound in the regolith (the only time NASA officially endorsed this use) for construction on the Moon. Interestingly, this report specifically used the picture in Figure 11.12 as a modern contrast to the original Bonestell picture shown in Plate 8. There was a clear intent to draw a historical line between the early work of the von Braun/Bonestell vision and the 1980s Reagan era vision of an equally robust space exploration program.

There was some justification for this expansive view of space development. In the early to mid 1980s, the Space Shuttle, while not reaching the promised flight rate, did some very interesting proof of concept missions. NASA had serviced the solar maximum mission, captured wayward GEO comsats whose upper stages had failed, and had deployed several commercial and government satellites. Several microgravity science missions had flown including the ESA provided Spacelab. NASA had started conceptual work on the second element of the von Braun-National Commission return to the Moon architecture, which was Space Station Freedom. Figure 11.13 shows the exuberance of the NASA astronaut corps, who, after a successful spacecraft servicing mission, gave an impromptu advertisement for NASA's version of commercial space:

Figure 11.13: NASA's 80s Dreams of Commercial Space

The National Space Commission was also a champion of commercial space and had several recommendations related to promoting activities such as commercial use of the Space Shuttle. They went even farther and advocated that NASA support private enterprise by giving incentives and favorable treatment to companies that developed private launch vehicles, space stations and in-space transportation systems.

The National Space Commission also threw in plans for space science, Earth observations, microgravity materials processing, and nuclear power for lunar bases and for NERVA type nuclear space tugs. In the spirit of leaving no interest group behind, the Commission also advocated educational programs, clean energy and terrestrial nuclear power systems, as well as research on ion propulsion and just about any other area supporting space that you could possibly think of.

The Commission was serious about advocating the use of resources from space, identifying legal impediments from UN treaties that should not be signed. Specifically they said:

> *Another example of pressures on the free use of space is the Moon Treaty negotiated in the United Nations, which would impose a generally more restrictive legal regime on the uses of space than existing agreements. In particular, the Treaty's provisions for the use of natural resources on other celestial bodies suggest a collectivized international regime analogous to the seabed mining regime in the Law of the Sea Treaty. Such a regime could seriously inhibit American enterprise in space. We therefore recommend that:* **The United States not become a party to the Moon Treaty.**[xiii]

The architecture for the return to the Moon advocated by the National Space Commission was very similar to the earlier work of the Lunar Bases and 21st Century Space Activities Symposium. The Shuttle and a heavy lift vehicle would provide the Earth launch capability needed. The Space Station would provide the assembly point and waystation on the way to the Moon. Beyond that, a Low Lunar Orbit space station would act as another waystation, a fuel depot, and safe haven for the lunar explorers. All in all, the National Commission report contained everything hoped for by the advocate community and this was partially responsible for its downfall. It simply cost more than the congress was willing to spend on it, something of a recurring theme in space. Also, an event happened as the report was being prepared that started us down the path that we are still on today.

On January 28th, 1986, as the National Commission on Space Report was being prepared, the Space Shuttle Challenger exploded after liftoff, taking the lives of seven astronauts, including the first civilian teacher in space. This explosion also took the life of many of the dreams for the Space Shuttle program and deflated the expansive ideas embodied in the Commission report.

NASA's credibility was severely damaged by the revelations of the Rogers Commission that investigated the causes of the Challenger accident. This, coupled with growing congressional opposition to the Republican administration of Reagan in its waning years, led to the National Commission on Space report virtually being ignored. A lot of the opposition was directed toward the very expansive nature of the recommendations in the report. The costs of this program were severely criticized as being unrealistic based upon growing concerns about the development costs of Space Station Freedom. This was in spite of Congress' own role in inflating those costs by delaying the program, holding back money and generally under funding the effort.

The Commission report languished until after the election of George Herbert Walker Bush in 1988. Bush reiterated his support for the Space Station and moved forward himself toward establishing an expansive space policy. This policy was somewhat less broad brushed than the one put forth by the Commission and more focused on the twin role of return to the Moon and on to Mars. This program, now known as the Space Exploration Initiative or SEI, begins the last chapter on lunar architecture development.

*   *   *   *   *

[i] http://hiwaay.net/~hal5/HAL5/L5-history.shtml

[ii] G. O'Neill, The *Colonization of Space*, Physics Today, 27, September, 1974, P. 32-40

[iii] K. Ehricke, *Lunar Industrialization and Settlement, The Development of a Polyglobal Civilization,* Lunar Bases and Space Activities in the 21st Century, Lunar and Planetary Institute Press, Houston, 1986, P. 830

[iv] http://ads.harvard.edu/books/lbsa/

[v] G. Woodcock, *Mission Modes for Lunar Basing*, Lunar Bases and Space Activities in the 21st Century, Lunar and Planetary Institute Press, Houston, 1986, P. 111-124

[vi] S. Hoffman, J Niehoff, *Preliminary Design of a Lunar Surface Research Base,* Lunar Bases and Space Activities in the 21st Century, Lunar and Planetary Institute Press, Houston, 1986, P. 71

[vii] W. Mendell, et al, *Strategies for a Lunar Base*, Lunar Bases and Space Activities in the 21st Century, Lunar and Planetary Institute Press, Houston, 1986, P. 57-68

[viii] N. Ruzic, *The Case for Going to the Moon*, G.P. Putnam Sons, New York, NY, 1965, P.5

[ix] http://lifesci3.arc.nasa.gov/spacesettlement/spaceresvol4/environment.html#Illustration

[x] Scientific Opportunities in the Human Exploration of Space (1997), National Academy of Sciences, Commission on Physical Sciences, Mathematics, and Applications, Space Studies Board, Committee on Exploration, The National Academies Press, 1997. P.86 Online at: http://books.nap.edu/books/0309060346/html/86.html#pagetop

[xi] Ibid, P. 88

[xii] Pioneering the Space Frontier, An Exciting Vision for the Next 50 Years in Space, The Report of the National Commission on Space, Bantam Books, May 1986

[xiii] Ibid, P. 166

Chapter 12

Lunar Exploration Architectures 1989-2003

SEI 90 Day Report, Augustine and Stafford Commissions, and NASA NeXt

The "Report of the 90-Day Study on Human Exploration of the Moon and Mars" was inspired by a declaration by President George Herbert Walker Bush on July 20, 1989, the 20th anniversary of the landing of Apollo 11. This declaration was the first step in a new plan to return the U.S. to the Moon. The key part of that declaration was:

> *"...a long-range continuing commitment. First, for the coming decade, for the 1990s, Space Station Freedom, our critical next step in all our space endeavors. And next, for the next century, back to the Moon, back to the future, and this time, back to stay. And then a journey into tomorrow, a journey to another planet, a manned mission to Mars. Each mission should and will lay the groundwork for the next..."*

This declaration began what is now known to history as the Space Exploration Initiative or SEI. This is what everyone had been waiting for, an executive commitment to return to the Moon and on to Mars! After 17 years and the death of the Apollo program, the President of the United States made exploration of the Moon and Mars a national commitment. Great hope was attached to this since it was very early in the Bush administration, which meant that he was attaching the prestige and the political capital of his office to support the effort. The presidential declaration also gave guidance to the exploration architecture for NASA. This was a modern version of the original Collier's von Braun architecture of coupling launch of cargo and humans to a space station where they would be assembled, tested, and sent to the Moon and then on to Mars. This architecture also built upon the work of the National Space Commission Report of three years earlier, but for a more focused mission to return to the Moon rather than the all encompassing approach advocated previously.

NASA soon created a task force to determine how to best carry out the declaration. The resulting document was the infamous 90-Day Study. In this study five reference architectures were studied in detail, all within the scope of the presidential directive. There were several commonalities within the five architectures, here described by NASA:

> *The five approaches presented reflect the President's strategy: First, Space Station Freedom, and next back to the Moon, and then a journey to Mars. The destination is, therefore, determined, and with that determination the general mission objectives and key program and supporting elements are defined. As a result, regardless of the*

implementation approach selected, heavy-lift launch vehicles, space-based transportation systems, surface vehicles, habitats, and support systems for living and working in an extraterrestrial environment are required.[i]

In this one paragraph were sown the seeds of hope and the seeds of doom for the program. The determination that a new heavy-lift launch vehicle was needed caused the costs to skyrocket. Adding to that burden, was the requirement that all of the architectures studied have high power nuclear systems designed for powering a lunar base. This system would be used to support operations during the 14 "day" night of the Moon. There was also the additional requirement to support development of hardware for a Mars mission. This would require reviving NASA's nuclear rocket program and a NERVA class nuclear thermal rocket. The high cost of development of all of these items put congress on the alert that this was going to be an expensive proposition.

Also, in a major architecture change that was to bode ill for space advocates was the deletion of *economic development* of the Moon as a rational, as Ehricke and the National Space Commission had advocated, to *"build a permanent outpost on the Moon to establish a human presence for science and exploration."*[ii] This seems to be the result of wanting to gain a greater acceptance from the scientific community. However, an inherent contradiction began developing as the NAS said specifically that science could **not** be the driving force behind human exploration. The report continued:

> *...at which human beings can learn to live and work productively in an extraterrestrial environment with increasing self-sufficiency, using local lunar resources to support the outpost. In this way, the lunar outpost will advance science and serve as a test-bed for validating critical mission systems, hardware, technologies, human capability and self-sufficiency, and operational techniques that can be applied to further exploration.*

With the above statement the lunar effort was downgraded from a base to an outpost, and resource development was moving into a support role for the further exploration of Mars, rather than an end unto itself.

The Five SEI Architectures Commonalities

Once the presidential declaration was made and major architectural choices were made, NASA began studying how to balance the different components, their development cost and schedule, to achieve their goals. To understand how the different components influenced the overall cost of the effort, the five SEI architectures each had a slightly different focus. For instance, one version was driven by schedule like the Apollo program of the 60s and another was paced to match the development of the major technology driven elements of the architecture.

Beyond the architectures for getting there, different objectives on the Moon and Mars were studied that required slightly different architectures and this heavily influenced the cost and schedule differences. If the utilization of the local resources of the moon (In Situ Resource Utilization or ISRU) were to be heavily emphasized rather than science, then the lunar and Mars outpost would look different. If budget was a problem, then the architecture could emphasize an approach that stretched out the spending according to how much money congress paid, or "go as you pay," a phrase coined by Norm Augustine's separate commission a year later. If congress became very stingy with the funding, the final scenario covered this possibility. Table 12.1 gives the different schedules for the five different architectures developed, listed as Reference A-E:

Milestones	Reference A	Reference B	Reference C	Reference D	Reference E
Moon					
Emplacement	1999-2004	1999-2004	1999-2004	2002-2007	2002-2007
Consolidation	2004-2009	2004-2007	2004-2008	2007-2012	2008-2013
Operations	2010à	2005	2005	2013	2014
Humans Arrive	2001	2001	2001	2004	2004
Permanent Habs	2002	2002	2002	2002	———
Local Built Habs	2005	2006	2007	2008	2011
Eight Crew	2006	2007	2007	2009	———
LUNOX Use	2010	2005	2005	2013	———
Farside Survey	2012	2008	2012	2015	2022
Steady State	2012	2008	2012	2015	———
Mars					
Emplacement	2015-2019	2010-2015	2015-2019	2017-2022	2024à
Consolidation	2020-2022	2015-2018	2020-2022	2022à	———
Operation	2022à	2018à	2022à	———	———
Humans Arrive	2016	2011	2016	2018	2016
Extended Stay	2018	2014	2018	2023	2027

Table 12.1: Key Dates for SEI Five Architectures

Common to all of the architectures except for "Reference E" is the use of lunar oxygen to help lower the costs of going to Mars. The study determined that 85% of all of the mass required for the initiative is in the form of oxygen so dramatic savings in logistics and launch infrastructure would be gained by deriving oxygen from the lunar regolith.[iii] Beyond the use of oxygen, all of the reference architectures for lunar and Mars exploration have science and exploration as their goal. A steady state outpost on the Moon and a base on Mars that would be continuously inhabited are common to all of the architectures except for "Reference E" as well.

An unstated commonality of all of the reference architectures is that a lunar outpost and a Mars base would look very much like government science bases on the continent of Antarctica. This would be a place to where up to eight scientists and other astronaut disciplines would study the Moon, develop oxygen extraction methods, and test technologies related to the eventual Mars mission. This would be the home of the few, the proud, NASA employees. Little place was ever considered in the SEI study for a more expansive program of resource exploitation, lunar manufacturing, or the growth of the base to include anything other than NASA's own people. The same was true of the Mars base as well.

Space Station Freedom

The baseline design of Space Station Freedom (SSF) was substantially modified from its 1989 look for the implementation of the SEI mission. This "new look" made SSF into a base for the assembly and servicing of exploration vehicles, fuel storage, and all the logistics necessary to make the space station into a "beachhead in the sky," for exploration. Figure 12.1 illustrates some of the enhancements that SSF would get as well as the "Extended" configuration:

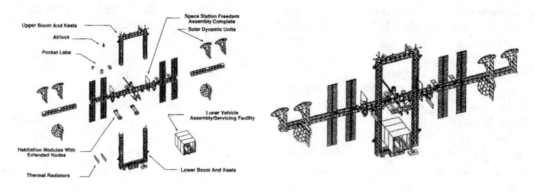

Figure 12.1: Space Station Freedom SEI Enhancements

The addition of the dual keel, power augmentation and hanger, would transform SSF from its microgravity science role into a powerhouse of on-orbit assembly and staging for moving out into the solar system.[iv] Freedom would be able to house more people in order to accomplish the assembly tasks and serve as a way station for people and cargo bound for the Moon and eventually on to Mars. Some of the original ideas for the space station as developed by Mendell and others at JSC for servicing GEO satellites were dropped as well. The GEO comsat industry wanted nothing to do with manned spaceflight after the Challenger disaster and the removal of all commercial spacecraft from the Shuttle manifest.

Heavy-lift Vehicles

The Space Shuttle could only lift about 18 metric tons (40,000 lbs), to the 28.5 degree inclination and altitude of Space Station Freedom. The Atlas II, Delta II, and the Titan IV would augment the Shuttle's capability. Of these, only the Titan IV had a comparable lift mass and volume to the Shuttle. The Delta II and Atlas II were brought back into production after the loss of Space Shuttle Challenger and the Titan was moved into accelerated production to make up the gap for national security payloads that would have otherwise flown on the Shuttle. All of these vehicles were classified as medium lift, having only a fraction of the capacity of the mighty Saturn V.

The Shuttle was supposed to be the U.S. principal launch vehicle for commercial and government payloads, but never lived up to its original promise, mostly due to funding starvation in its early days of development. To further augment the Shuttle, which was now to launch only humans and high value cargo to the station, and the newly revived ELVs, a Shuttle "C," or cargo version, was proposed as the first new start for launch vehicle development under the SEI program. The first versions of the Shuttle C (shown in Figure 12.2) would lift between 61 or 80 metric tons to the orbit of SSF depending upon whether two or three Shuttle main engines were used.

Figure 12.2: Shuttle C, A Shuttle Derived Heavy-lifter (Mockup at MSFC)

The cargo bay of the Shuttle C would be stretched from 60 X 15 feet to 85 X 15 feet. Lift mass would be increased by eliminating the crew cabin, wings and thermal control system tiles that protect the crewed shuttle on reentry into the Earth's atmosphere. The Shuttle C would be used to lift the lunar tug, the lunar lander and other heavy payloads. This lowered, but did not eliminate, the requirement for on-orbit assembly since most of the functionality of the hardware would be tested on the ground before launch. In Figure 12.2 above, mock-ups of a Space Station Freedom long module, a Space Station node and some truss hardware are shown in the cargo bay and would be launched in one mission.

After Space Station Freedom was built with a combination of Space Shuttle and Shuttle C hardware, lunar mission hardware would also be sent up to the station to be assembled. Figure 12.3 shows some of the SEI lunar hardware.

These lunar bound vehicles would be carried up in pieces inside of the Shuttle C. The fuel tanks, landing legs, aerobrake and cargo would be added at the station. This would be accomplished with a mixture of human and robotic assistants. The system would be tested in the hanger at station as well as fueled and provisioned, taking a page from the original von Braun ideas as outlined in Collier's (Plate 9). The station would carry large tanks that would be filled by incoming Shuttle C or expendable vehicle tankers, or the lunar bound vehicles could be fueled directly by the tankers. The vehicle shown in Figure 12.3 illustrates the assembled system. Because of its size, there is no way that the fully assembled vehicle could fit into the 5 meter diameter of the Shuttle C cargo bay.

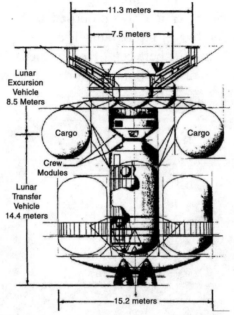

Figure 12.3: Station Assembled Lunar Vehicles (All SEI Graphics Courtesy NASA)

Later, as the program moved forward, the Advanced Launch System (ALS), a joint NASA USAF Shuttle derived heavy-lifter, would be able to lift payloads comparable to the Saturn V (120-140 tons) for habitation modules, resource utilization payloads and most importantly, fuel. Later, this would be augmented by an even heavier vehicle that could launch upwards of half a million pounds, or 250 tons, to orbit. This would be principally to support a Mars mission, but would be an optional way to put the lunar elements into orbit without relying on on-orbit assembly. Figure 12.4 gives a comparison between the Shuttle C and the Advanced Launch System (ALS) for exploration.

The advocates of the heavy-lift option touted the ability to assemble most of the lunar payload on the ground and fit it inside the much larger fairing (a fairing is the payload container that protects the payload from rain and from damage due to wind loads during launch) of the ALS booster. Also, it could carry the weight of the Lunar Excursion Vehicle (LEV) and also carry the Lunar Transfer Vehicle (LTV) into orbit and boost it to space station in one launch. The total weight of the LEV and LTV fully fueled would have been 129.8 metric tons. Fuel would be launched separately on a Shuttle C or another ALS vehicle. There was even the possibility that the ALS derived vehicle could carry the whole payload to the Moon, bypassing the space station.

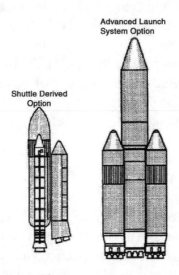

Figure 12.4: Shuttle C & ALS

Common Mission Architecture

In all of the SEI reference missions studied, the architecture for returning to the Moon looks like the illustration in Figure 12.5 below from the "Report of the 90 Day Study."[v]

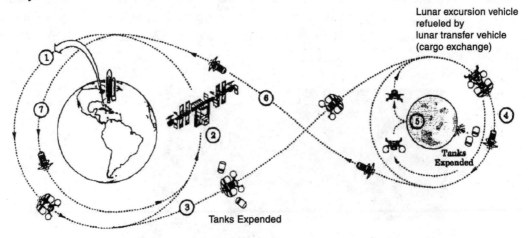

Figure 12.5: SEI EOR/LOR Mission Architecture

Payload delivered to Space Station Freedom
Lunar Transfer vehicle mated with payload at Freedom
Trans-lunar phase with lunar transfer vehicle
Lunar transfer vehicle rendezvous with lunar excursion vehicle from Moon
Excursion vehicle returns to Moon with payload

Trans-Earth phase with transfer vehicle
Transfer vehicle aerobrake maneuver and return to Freedom

One difference between this architecture and previous architectures from the 80s was the deletion of the low lunar orbit space station, no longer considered necessary with the use of the automated Lunar Transfer Vehicle. One throwaway aspect of this architecture is the use of expendable tanks on the LTV. Using expendable tanks allows a 10% gain in payload, but entails the necessity of boosting new tanks up on every mission. The rest of the mission scenario was similar to those discussed in Chapter 11 with regard to using the aerobrake to slow the returning spacecraft down and return it to Space Station Freedom.

Lunar Base Implementation

Other technological aspects were also similar to the previous vision as laid out by the National Commission on space. The lunar outpost would use nuclear power, the mission architecture shown above in Figure 12.5 would be stretched with added strap-on fuel tanks to go to Mars and establish a base there. The use of oxygen for propulsion was central to most of the design reference missions and would be used for providing fuel for the LTV returning to the Earth as well as fuel for any Mars-bound vehicle.

The lander used for the effort would be quite capable, with the ability to carry several tons of cargo including modules, rovers and other system components. By using on-orbit assembly, the size and design limitations that come with being confined within a launch vehicle fairing go away. This is an incredible advantage that even today is not fully appreciated by the aerospace community. Figure 12.6 shows the LEV on the surface of the Moon:

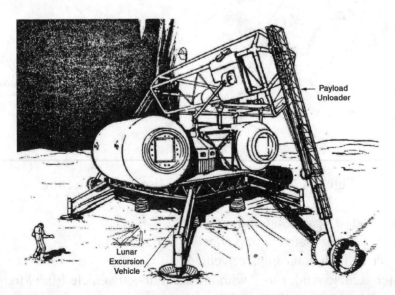

Figure 12.6: Lunar Excursion Vehicle with Payload and Unloading Crane

The lander shown above is the same one shown docked to the LTV in Figure 12.3. To understand the power of on-orbit assembly, it is crucial to understand that weight is less of a problem with a lunar cargo carrying lander than is available volume. The measurement of the distance from the outside of the module on the left side of the lander to the module on the right side of the lander exceeds 17 meters (55.7 feet)! There is no practical way to build a fairing with that diameter for a launch vehicle. To transport cargo of this size any other way would require multiple flights carrying smaller loads. What this means is that if you use a heavy-lift vehicle the total number of flights to the Moon or Mars actually increases over what would be the case if you transported the parts to the space station and assembled them into a larger unit.

Also, having a servicing center in LEO allows for extensive testing before a flight to the Moon or Mars is begun. Beyond this, the existence of the servicing center, where fuel can be stockpiled, servicing of the returning vehicle can be accomplished, and a biological quarantine of the Earth returning explorers, is important. Such a facility supports reusability of the lunar and Mars systems, which considerably lowers cost. Also, using ISS as a quarantine facility off of the earth greatly improves the ability to isolate any potential pathogen picked up by a human crew from Mars.

For power generation, the very first flights would use solar power, but after that, a nuclear power source generating a minimum of 100 kilowatts would be landed and put into operation. Figure 12.7 shows this setup:

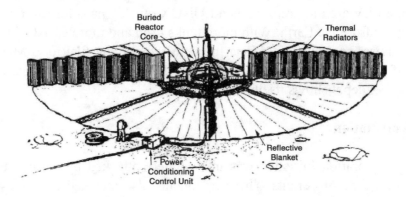

Figure 12.7: 100 Kilowatt Nuclear Reactor for Lunar Base Power

This reactor was required to power the lunar oxygen plant needed to cut the logistics cost of launching large amounts of oxygen from Earth to provide propellant to the LTV and LEV. It would also power the base during the long 14 "day" lunar night.

Another facet of the SEI architecture that was inherited from prior studies and recommendations is the use of lunar oxygen, or LUNOX. This is included in all but one of the architectures studied. The use of LUNOX again dramatically reduces the mass needed to lift loads from the Earth, something that calls into question the need for large launch vehicles after a robust LUNOX infrastructure is in place. Figure 12.8 shows a prototype LUNOX manufacturing plant:

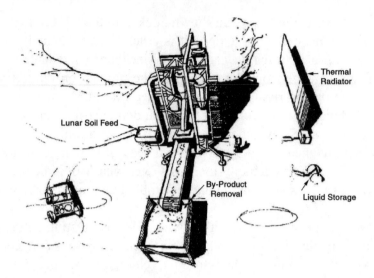

Figure 12.8: LUNOX Prototype Plant

In the scenario that uses the LUNOX plant, the LEV lands with its fuel tanks empty and can carry extra fuel tanks as payload. These are the same tanks as are used on the LTV. The LUNOX is loaded and then these fully fueled tanks are used to replace the expended tanks on the LTV in orbit. This provides a wide margin of fuel for the return trip and a partial or full load for the next trip back from LEO to the Moon. This could save as much as a full heavy-lift launcher per round trip mission to the Moon. If these tanks and the LEV are fully reused, then LUNOX quickly pays for itself in the saving of launch mass from the Earth. With more tanks used and more LUNOX it should be possible to amass enough fuel in lunar orbit to provide the fuel for a Mars mission, especially if lunar water is used. This is where the value of the low lunar orbit space station as a fuel depot shows itself.

SEI Implementation

The Space Exploration Initiative was immediately attacked by various foes as too expensive in a time of deficits. These opponents also attacked it as a $400 billion dollar "Moondoggle". NASA did not help this criticism since the plan over 35-40 years would cost something close to this amount, not only for the lunar missions, but included Mars, space science and all of NASA's other activities.

NASA was also being legitimately criticized at the time by congress and the aerospace press for an inability to control costs and perform technically on large programs such as Space Station Freedom, the Space Shuttle program and unmanned missions. The Space Shuttle was also beset by technical problems such as a hydrogen leak that kept the Shuttle grounded for months, casting doubts on the ability of NASA's workforce to be able to execute a supremely technical enterprise as recommended by the 90 day study. In yet another blow to the technical credibility of the agency, the Hubble Space Telescope, launched in early 1990, had a problem equivalent to having the wrong prescription for glasses that rendered the telescope useless for its intended role as the world's highest resolution telescope.

Another disturbing element that gave credence to cost and personnel concerns was the request by NASA in the 90 day study for *"... a significant augmentation of civil service positions to support the Human Exploration Initiative. These positions will provide the necessary technical expertise, program management, and administrative support to meet the objectives of the Initiative."[vi]* This came at exactly the wrong time because the federal government, groaning under deficits, was looking for ways to cut the government payroll, not expand it. As the plan began to stall, yet another commission was appointed, led by defense contractor Martin Marietta's former CEO, Norm Augustine, to address the problems and provide recommendations to the President on the direction that NASA should go.

The Augustine Commission

The "Report of the Advisory Committee On the Future of the U.S. Space Program," more commonly know as the Augustine report, was commissioned by the White House as a response to concerns about NASA's ability to execute its mission. The Augustine report would result in a complete change in the way that NASA worked and would dramatically affect the plans of the Space Exploration Initiative.

In its executive summary, the Augustine Commission began to detail both the recent failures of NASA and the successes. Only a few months before, the Voyager II spacecraft had performed a flyby of Neptune with an error in navigation of less than 19 miles out of the more than one billion miles that it had traveled, a feat of navigation never before attained. The Magellan Venus orbital mapper had recently completed the first medium resolution radar map of the veiled planet. The Commission also indicated that in the past, NASA had lost far more spacecraft during the Apollo era than was the case by 1990. With this being said, the Augustine Commission said something very baffling.

> **Concerns:** *Nonetheless, given the cost of space activities, in both financial and human terms, and their profound impact on America's prestige throughout the world, no goal short of perfection is acceptable. The Committee finds that there are a number of concerns about the civil space program and NASA, which are deserving of attention.[vii]*

No goal short of perfection? This is indeed a high standard to uphold. What is baffling is that the reason for the goal was the *"profound impact on America's prestige throughout the world."* This was an extension of the advertising reason for going to space first used to justify the Apollo program. The Augustine Commission's demand for perfection would lead to a dramatically curtailed NASA exploration plan and greatly increased costs, exactly the things that their commission was supposed to help mitigate.

To say that the Augustine Commission report was a train wreck to the direction that NASA had been going for over 30 years is an understatement. As a result of the commission's work, NASA was asked to completely abandon the space station as a transportation node and assembly area for manned spaceflight beyond Earth orbit. The Commission expressed the value of the space program in the following way:

Yet perhaps the most important space benefit of all is intangible — the uplifting of spirits and human pride in response to truly great accomplishments — whether they be the sight of a single human orbiting freely around the Earth at 18,000 miles per hour, or a picture of Uranus' moon Miranda transmitted 1.7 billion miles through space, and taking some 2-1/2 hours merely to arrive at our listening stations even when traveling literally at the speed of light. Such accomplishments have served to unite our nation, hold our attention, and inspire us all, particularly our youth, as few other events have done in the history of our nation or even the world.

This statement, along with the previous one about prestige brings the space program back full circle to the justifications espoused at the beginning of the Apollo era. Gone are all of the justifications that were built up during the visionary years of the 70s and 80s of bringing actual tangible value to the American people by the actual development of the resources of the Moon and extending civilization there. The intangible follows the tangible and not vice versa. This mode of thought is reflected in their recommendations and conclusions enumerated both in the executive summary and in the body of the document.

The whole priority of the Agency was changed from an outward focus on exploration and the development of the resources of the solar system, to a two-tiered science-based focus. The Augustine Commission expressed the first and most important area of focus:

*"It is our belief that the **space science program** warrants **highest priority** for funding. It, in our judgment, ranks **above space stations, aerospace planes, manned missions to the planets**, and many other major pursuits, which often receive greater visibility. It is this endeavor in science that enables basic discovery and understanding, that uncovers the fundamental knowledge of our own planet to improve the quality of life for all people on Earth, and that stimulates the education of the scientists needed for the future. Science gives vision, imagination, and direction to the space program, and as such should be vigorously protected and permitted to grow, holding at or somewhat above its present fraction of NASA's budget even as the overall space budget grows."[viii]*

This was a complete reprioritization of the nation's space program. In the bold type above, the space science program became ascendant as the NASA "raison d'etre," or reason for being. No longer were space stations, manned missions to planets and space exploration to be NASA's primary focus. The following further reinforced this:

*"Having thus established the **science activity** as the **fulcrum** of the entire civil space effort, we would then recommend the "mission-oriented" portion of the program be designed to support two major undertakings: a Mission TO planet Earth and a Mission FROM Planet Earth. Both, we believe, are of considerable importance. The Mission to Planet Earth, as we would define it, is the undertaking that in fact brings space down to Earth — addressing critical, everyday problems, which affect all the Earth's peoples. While we emphasize the need for a balanced space program, it is the Mission to Planet Earth, which **connotes some degree of urgency**. Mission to Planet Earth, as we would define it, comprises a series of Earth-observing satellites, probes and related instruments, and a complementary data handling system aimed at producing a much clearer understanding of global climate change and the impact of human activities on Earth's biosphere. This effort will provide us with a much better understanding of our environment, how we may be affecting it, and what might be done to restore it."*

To me and in the context of this book, the above statement means that NASA should be studying how we are affecting the planet rather than how to save it! This reflected a mindset of the scientific component of this commission that was philosophically aligned to the "Limits to Growth," mentality.

This was such a complete turnaround of the agency's focus that it actually brought NASA back to where it was in the early 70s after the death of the Apollo program when Earth Resources (LandSat) and other Earth oriented science programs dominated NASA. In the Augustine Report worldview, space science became the "fulcrum" of NASA efforts around which everything had to revolve. The primary mission of NASA would become the "Mission to Planet Earth," and the focus would become climate change and it was the commission members profound belief that this focus was the appropriate one for the agency.

It was the Commission's opinion that, if the path that they laid out was not taken, the space program would merely "drift" through the coming decade. This "drift" was caused, in my opinion, by the failure of NASA to come up with an exploration agenda that was cost effective and included the American people as more than onlookers. This call, to have science be the fulcrum, created a gulf between the Commission's study and every other study that had gone before, and ignored the National Academy of Sciences opinion that exploration of the Moon and Mars could not be justified on scientific grounds. The other focus of the Commission's study, the "intangible" value, is both unappealing and contradictory. It is unappealing because simply sending astronauts to the space station to make circles around the Earth has little intangible value for inspiration and education, and is contradictory because, if robots, as the commission suggested, are truly the best at doing exploration, then why have humans in space at all?

The Augustine Commission hammered more nails into the coffin of exploration with the following statement:

> *"But if there is to be a manned space undertaking, what should it be? Surely the goal is not merely to provide routine transportation of cargo to and from space. In this regard, we share the view of the President that the long-term magnet for the manned space program is the planet Mars — the human exploration of Mars, to be specific. It needs to be stated straightforwardly that such an undertaking probably must be justified largely on the basis of **intangibles** — the desire to explore, to learn about one's surroundings, to challenge the unknown and to find what is to be found. Surely such an endeavor must be preceded by further unmanned visits, and by taking certain important steps along the way, including returning for extended periods to the Moon in order to refine our hardware and procedures and to develop the skills and technologies required for long term planetary living."*

In this new scheme of operation for NASA, the Moon becomes a sideshow and not a primary goal for the space program. In this one paragraph ended any hope for Ehricke's Selenopolis, the use of lunar resources for O'Neill colonies, solar power satellites, or trips to Mars and beyond. It also killed any thought of the industrialization of the Moon and the use of its resources to benefit the people of the Earth. Even the whole purpose of going to Mars was reduced to "intangibles," "the desire to explore," and other similar reasons that continue to foster a government space mentality with little hope of participation by the wider population.

Space is not an advertising campaign! This is the antithesis of common sense and the aspirations of generations of exploration and development of our own planet. The manned or unmanned exploration of space is not worth the treasure that it costs if the results are merely to "uplift the spirit," whatever that means. This attitude is what killed the Apollo era. There was a picture shown in the New York Times in 1970. It was the famous picture of Buzz Aldrin on the Moon, but modified by some witty person. Buzz was holding a sign that read, "So What?" Space exploration by humans must make a connection with and make contributions to solving our problems today on the Earth or it serves little purpose. That is what is truly uplifting about space, its potential for our future.

The lunar program was reduced to that of a "possible" outpost as shown in recommendation seven of the Augustine Commission's report:

Recommendation 7: That technology be pursued which will enable a permanent, possible man-tended outpost to be established on the Moon for the purposes of exploration and for the development of the experience base required for the eventual human exploration of Mars...

The Commission's report represented a dramatic shift away from the practical aspects of manned exploration with a polite nod toward Mars. In this, however, they recommended changes that denied even the possibility of going to Mars by removing the space station as an assembly area and waypoint for exploration. Their recommendation related to the space station was:

...we doubt that the Space Station will be essential as a transportation node — certainly not for many years. However, the Space Station is deemed essential as a life sciences laboratory, for there is simply no Earth-bound substitute. The Space Station is a critical next step if the U.S. is to have a manned space program in the future... ...Given these conclusions, we believe the justifying objectives of the Space Station Freedom should be reduced to two: primarily life sciences, and secondarily microgravity experimentation...

They went further to recommend that the space station be redesigned to fit the new mission. On-orbit assembly would be cut back and most of the space station systems built, integrated and tested on the ground. They also recommended that reliance on the Space Shuttle be cut back and assisted by the development of a new heavy-lift vehicle for station and future human space activities. Both of these recommendations were based on cost considerations and a reduction of Extra Vehicular Activities (EVA) or spacewalks. What they did not realize is that this was a contradictory set of recommendations in the event the heavy-lift vehicle wasn't built, which is what actually happened. The Shuttle C shown in figure 12.2 never got beyond the mockup stage.

Like many of the other commissions had in the past, the Augustine report recommended that nuclear power sources be developed, technology development be fully funded for new and innovative launch vehicles, lower cost robotic spacecraft, and a new generation of NASA engineers and scientists be hired to replace those nearing retirement. However, these recommendations were mostly ignored except for the low cost robotic craft. This became the Dan Goldin cry of "faster, cheaper, better," that did increase the number of unmanned spacecraft sent. This was one of the few positive aspects of the report that NASA adopted.

The Ramifications of the Augustine Report

It is curious that the Augustine report directly contradicted President Bush's Space Exploration Initiative. The only way to understand this is to recognize that the members of this commission were mostly from the space science community and did not share the vision of either the President or the commissions, studies and programs that came before. For the last 20 years there has been a growing tension between the space science program and the human spaceflight program. There have been many arguments about how manned space takes money from unmanned space, or how robotic spacecraft can do the exploration job better than humans. This is evident in the Augustine Commission report.

The Augustine Commission report would have sat on the shelf, as many of the past reports had, if it were not for one reason. In 1992, President George Bush appointed Daniel Goldin as NASA Administrator to shake up the organization. Goldin used the Augustine report as a virtual bible in his mission to remake the organization. But before the Goldin era started, one more commission report, the Stafford Commission, was to have its influence on efforts to return to the Moon.

The Stafford Commission

From late 1990 through May 1991, Major General Thomas Stafford (retired), a former Apollo era astronaut, led a "Synthesis Group," that was tasked by the President and Vice President, to synthesize the various commissions and reports concerned with NASA's future into some coherent form that was compatible with the vision of returning to the Moon as outlined by the president on July 20, 1989. This included most of the reports discussed in this and the previous chapter. The synthesis activity also included taking input from the public and the membership of the American Institute of Aeronautics and Astronautics (AIAA).

The Stafford Report and its recommendations are in accordance with the studies and commissions that came before, but it was also heavily influenced by the Augustine Commission report. Here is the final paragraph of the forward to the report:

> *Space is clearly our most challenging frontier. Enroute to Mars, we will explore the Moon, advance Earth sciences, and develop new, innovative technologies. We will tap lunar, Martian, and solar energy resources as we explore the heights of human talent and ability. Along the way, America's drive, initiative, ingenuity and technology—all those things that have made our nation the most successful society on Earth—will propel us toward a future of peace, strength and prosperity. The challenge is before us. This report shows how we begin.*[ix]

So, in concept, the Synthesis Group report is an echo of the previous works, but with considerable influence from the radical change in direction advocated by the Augustine Commission. The Synthesis Group report had three basic architectures based upon a group of exploration waypoints guided by a National Space Vision. The "Moon Waypoints" and "National Space Visions" listed below.[x]

Moon Waypoints

Lunar Exploration
Preparation for Mars
Habitation
Lunar Based Observation
Fuels
Energy to Earth
Asteroids Waypoint
Mars Waypoint

National Space Visions

> Increase our knowledge of our solar system and beyond
> Rejuvenate interest in science and engineering
> Refocus U.S. position in world leadership (from military to economic and scientific
> Develop technology with terrestrial application
> Facilitate further space exploration and commercialization

The three architectures proposed were similar to one another and designed to reflect the vision and waypoints. These architectures are discussed in the next section.

The Stafford Architectures

The three architectures developed by the Synthesis Group were:

> Emphasis on Exploration and Science
> Emphasis on Human Presence
> Emphasis on Space Resource Development

The three architectures were all based on the ultimate goal of going to Mars. This shift in focus, as compared with the National Commission on Space five years before, is apparent from the amount of text dedicated to talking about Mars. Less than half of the National Commission on Space report dealt with Mars. In the Synthesis Group report, this had decisively shifted toward a Mars mission.

Another major shift that seems to have resulted as an influence of the Augustine Commission report was a move away from on-orbit assembly toward the use of heavy-lift vehicles and rendezvous in orbit for the transfer of fuel and other cargo. The report doesn't even mention the Space Station in the context of exploration. Following is the listing of the system components and technical strategies that underpinned the three exploration architectures:

Technical Strategies

> Develop a heavy-lift launch capability
> Limit on-orbit assembly
> Develop nuclear technologies
> Use the Moon as a test bed in preparation for Mars
> Use common systems and operations between the Moon and Mars
> Use a complementary mix of human and robotic resources
> Emphasize technologies with terrestrial applications

The mission architecture chosen as the guide for all of the lunar oriented missions was essentially the old Apollo style Lunar Orbit Rendezvous (LOR). However, the commission recommended the development of a heavy-lift vehicle with more than twice the lift mass of the Saturn V. This would have been a system equivalent to the Nova 8 described previously, but enhanced by at least partial reusability. This super-booster would also be the baseline for Mars missions.[xi] Ominously, the report states that:

Vice president Quayle asked that we investigate options to accomplish America's exploration goals faster, cheaper, safer, and better. This investigation has led to the very clear conclusion that to achieve these goals, the utilization of a heavy-lift launch vehicle having a capability to launch 250 metric tons to orbit is required.

This is ominous in that everything else in the whole plan is dependent upon the development of the heavy-lift vehicle. This is an example of what I call the "Apollo mentality." This mentality is defined as "we went to the Moon this way before, this is the only way to do it now." On-orbit assembly was nixed as not contributing to this goal of "faster, cheaper, and better." While the decision was in keeping with the findings of the Augustine Commission, no real comparison was ever made of the value of super heavy boosters versus on-orbit assembly, it was simply assumed to be true.

The first Stafford architecture was focused on lunar science and Mars preparatory activities. The Moon's primary value in the first architecture was for its science purposes. This idea was in direct conflict with the National Academy of Sciences again, since the science justification for the Moon was specifically not considered of high priority relative to other science activities.

The second Stafford architecture was science justified for both the lunar and Mars phases. There would be a base, much like an Antarctic base at the Earth's south pole. Science would predominate the lunar base with optical, infrared and radio telescopes emplaced. Mars supporting activities for the base were to be developed as well, such as closed life support and food growth activities.[xii] No In Situ Resource Utilization was to be used at all in this scenario. However, this architecture had as an option, a new idea, not part of any of the previous works, concerning an option for a mission to a Near Earth Asteroid! This option was placed in there as an easier task than going directly to Mars and landing since the energy required would be much less. For Mars, In situ resources of oxygen and possibly water would be used to lower costs and improve the capabilities of the exploration effort there.

The third Stafford architecture more closely resembled prior studies and was focused, at least initially, on lunar activities in science, exploration as well as resource extraction and utilization activities. The first return mission by humans would be executed by a six-person crew: five landing and one staying in orbit as in the Apollo days. This mission would be to conduct an initial survey for a location to set up the lunar base. An illustration of the lunar base is shown in figure 12.9:

Figure 12.9: Lunar Base the Stafford Report Version (Picture Courtesy NASA)

This first mission would also bring a nuclear power supply, science telescope and a solar flare warning system. This would be followed by cargo flights, bringing robotic hardware, the habitat and other essentials to begin the operation of a base. The next wave of explorers would bring with them more science gear and new cargo flights would bring ISRU units, food growth units and other hardware to begin the process of building up a self-sustaining base. By 2009, there would be a total of 18 people on the Moon dedicated to science, exploration and the continuing build up of ISRU capability. The ultimate focus would still be on Mars and the applicability of the lunar base to Mars exploration. This architecture still more closely resembles a government Antarctic base, staffed by government scientists and engineers, rather than any participation by a wider cross section of the American people. The American people were not impressed. This lack of a wider participation that began with SEI and was the norm for Augustine and Stafford helped to create a critical disconnect between the ones getting to have the fun and the ones paying the bill.

Stafford Bites the Dust

By presidential election time in 1992, SEI was dead. It was a victim of a mixture of reduced confidence in NASA to carry out large space projects on budget and schedule, a growing deficit due to the Savings and Loan debacle, and a congress unwilling to fund anything space related that was authored by George Bush. I was personally involved and affected by this. My team at the University of Alabama in Huntsville had won a contract with NASA, under Boeing's leadership, for the Lunar Resource Mapper spacecraft. NASA awarded the contract to Boeing and our group was to build a microsatellite that would be deployed from the main spacecraft to act as a beacon for mapping the gravitational field of the Moon. Congress took the money back under a little known "rescission" act as a way to stop any discoveries from the spacecraft from positively affecting the politics of support, effectively keeping "the camel's nose from under the tent."

The congress did have some valid reasons for rejecting SEI. The heavy-lift launch vehicle that the Synthesis Group declared as absolutely necessary would have cost tens of billions of dollars to develop. The Group's reason for developing the heavy-lift launch vehicle was to do the mission faster, cheaper, and better, but there was little evidence to support this argument. Von Braun once said that the only difference between a small and a large launch vehicle is a lot of money. NASA was also increasingly incapable of building any large project on time or anywhere near the amount of money budgeted. Congress used the utter failure of NASA to accurately contain the cost of space station as the evidence of their concerns. Indeed at this time NASA was under congressional direction to redesign the station to lower its costs, after the initial $8 billion dollar price tag had ballooned to several times that number.

The Synthesis Group also ignored the existence of the Space Station, rendering any value from on-orbit experience moot. As a result of the Augustine Commission, there seemed to be a retrenchment from astronauts being outside in space suits as a safety measure and due to the assumed inability of astronauts to effectively work in the space environment. This argument was undercut by the work of astronauts for the Intelsat 6 rescue in 1992, the dramatic and successful fixing of the nearly blind Hubble Space Telescope, and the more recent work constructing the International Space Station. There was no real justification for ignoring space station. To do the space station, which was already costing billions of dollars per year was another black mark against the Stafford Report and ultimately the entire SEI concept. SEI's fate was sealed with the Bush 41 administration in 1993 when president Clinton came into office. For most of the next 8 years space exploration would not be a focus at NASA.

The Post SEI Era, The Goldin Age

In 1993, NASA Administrator, Dan Goldin, appointed by Bush and kept on board by Clinton and Gore, was enthusiastically working to "reinvent" NASA as the effort to reorganize government was called. Budgets were cut, most of the last of the Apollo era NASA cadre were bought out and retired early, and NASA's plans were de-scoped to match the recommendations of the Augustine commission. Earth sciences were bolstered with the enthusiastic support of Vice President Gore. Technology development was increased, but in a futile way with a march of failed X-vehicles that promised "Cheap Access to Space" or CATS, but delivered nothing but rusting hardware, millions of dead trees sacrificed for paper reports, and hard disk drives filled with studies.

After the Clinton administration came into office, all talk of the Moon was forbidden. NASA took to heart the Augustine Commission about microgravity research and life sciences while cost overruns on the now renamed International Space Station (ISS) continued to consume the NASA manned spaceflight budget. Goldin took the money allocated by congress for the microgravity research and applied it to construction costs. Since Microgravity research was led by MSFC, this gave ISS lead center JSC the power to finally triumph over their rival MSFC in the derby to control ISS. This victory did not stop the overruns and led to the ouster of the legendary JSC Director, George Abbey, This was one of the first firings at NASA after the inauguration of George W. Bush in 2001.

Some faint light began to show again in the late 1990s when NASA Code S (Science) and Code M (Spaceflight) began new studies regarding various architectures for returning to the Moon and trips to Mars. These stayed far enough under the radar to not draw any congressional attention and will be discussed in the next chapter.

There is a common thread of high cost that runs through all of the architectures that I have written about here in the last two chapters that helped to guarantee the demise of these plans. A second thread of "science over everything" began to appear in the National Academy of Sciences report as well as the Augustine and Stafford reports. This new thread also had a negative impact on the salability of the overall plans. As can be easily seen, there was a huge amount of passion, work, hope and dreams woven into all of these plans, but ultimately, they all came to a futile end in the waning years of the 20th century. It has been almost 32 years since mankind last set foot on the Moon. This has to change, but it has to be changed for reasons associated with making life better here on the Earth! In the next chapter, this will be addressed and a vision will be described that can take us back to the Moon, for profit, and for the people.

*   *   *   *   *

[i] Report of the 90 Day Study on Human Exploration and Development of the Moon and Mars, National Aeronautics and Space Administration, November 1989, P. 1-1

[ii] Ibid, P. 3-1

[iii] Ibid, P. 8-10

[iv] K. Brender, AIAA 90-3758, *Space Station Freedom Flight Operations in Support of Exploration Missions*, NASA Langley Research Center, Hampton, VA, AIAA Space Programs and Technologies Conference, September 25-28, 1990, Huntsville, AL P. 8

[v] 90 Day Report, P. 3-13

[vi] Ibid, P. 1-7,8

[vii] *Report of the Advisory Committee On the Future of the U.S. Space Program*, http://history.nasa.gov/augustine/racfup2.htm (Augustine Commission Online),

[viii] Ibid, Executive Summary

[ix] *Report of the Synthesis Group on America's Space Exploration Initiative*, Prepared for the Vice President, Dan Quayle, Chairman, National Space Council, The White House, Washington, D.C. 20500, May 1991, P. iv

[x] Ibid, P. 11

[xi] Ibid, P. 31

[xii] Ibid, P. 43

Chapter 13:

Return to the Moon, The New Millennium

Developing the Vision

In the first 8 chapters of this book, I dealt with the question of why we want to go back to the Moon. Returning to the Moon involves developing the vision of *why* we want to return to the Moon. There is an old biblical saying, "Where there is no vision, the people perish." At no point in our history is this more true than today. The definition of vision in this context is, "sense of purpose." We must have a sense of purpose related to why we are returning to the Moon. The purpose, the "vision," must be to use the Moon's material resources to help us transcend the limits to growth and improve our lives here on the Earth. The limited extent of resources on the earth and the pollution resulting from our use of them are the greatest problems that confront mankind's first global civilization and its six billion people (nine billion by 2050). If we do not solve this problem, then the billions who live today will be reduced to millions tomorrow. Those who say that technology is not the solution to the material problems that face us are simply wrong and I hope that I have presented enough evidence to support this contention.

No other reason for returning to the Moon makes sense in the context of the problems that face us today. Happily, the resources derived from metal asteroids impacted on the Moon give us an environmentally responsible alternative to the non-solution of trading CO_2 credits with third world countries that the Kyoto Accord proposes. We need solutions that bring results in order to solve our resource and energy problems. The PGMs that form the foundation of the hydrogen economy food chain bring this solution, because, without inexpensive PGMs, fuel cells are impractical and the hydrogen economy implausible. We know from the statistics that we have presented here that PGM resources are limited on the Earth, and costly to extract.

We already know from Apollo samples that PGM resources do exist in diffuse form on the Moon. These exist in the microscopic nickel/iron fragments that make up 0.1 to 1% of typical Regolith samples brought back from the Apollo landing sites. [i] The central hypothesis here is that, due to the dynamics of how impacts occur, there is a high probability of large quantities of concentrated resources; enough to enable the hydrogen economy, along with enough "waste" cobalt, nickel and iron to fuel the development of the solar system.

The measure of this value must be to deliver lower cost platinum while generating a profit for the lunar economy. PGMs are the highest cost item for fuel cells and the principal barrier to their mass implementation to displace internal combustion engines. Lower cost platinum would make fuel cells economically competitive with the internal combustion engine. This means extracting the PGMs for a cost that results in a price that is well below that of today's market. This will allow the breathing room that our civilization needs to find alternatives to oil and move toward nuclear fusion power to make the vast quantities of hydrogen to power our global transportation infrastructure for the conceivable future.

It is possible that PGMs alone will not be quite enough by themselves to make this economically feasible but it will be the activity that generates revenue and local resources to enable a lunar settlement, so that is where we begin. Other activities that have a higher profit margin will be enabled by the activity of PGM extraction and the infrastructure that is created for this activity.

Implementing the Vision

In chapter 9, I reviewed all of the past lunar missions to address their contributions toward initial exploration of the Moon. These missions, while fruitful in themselves, were of limited utility for identifying PGM concentrations. The unmanned missions did give us a tremendous amount of information on the global resource potential of the Moon and the Apollo missions gave invaluable ground truth data to confirm and calibrate the orbital data. We now know the amount of oxygen, iron, titanium, aluminum, silicon and trace volatiles on the lunar surface. The Clementine and Lunar Prospector missions of the 1990s gave us our tantalizing glimpses of the possibility of water in the permanently shadowed regions of the lunar poles. When put together, these data sets constitute a great start, but still do not give us all of the information that we need in order to go back to the Moon to develop its resources.

In chapters ten through twelve, I went through all of the previous architectures: the Collier's imaginative vision of the 50s, the cold war ad campaign that was Apollo, the wilderness years of the 70s, the exuberance of the 80s and the dashed hopes and dreams of the 90s. All of the historical architectures and visions have one thing in common: *They failed*! Even Apollo is a failure when measured against a practical metric of enabling the development of the Moon and its resources.

What Went Wrong and Where Do We Go From Here?

All of these architectures failed for a common reason, they did not address issues of central importance to the American people. Apollo was a success only in strict terms of its role in fighting the cold war, but it failed to extend its purpose beyond that narrow goal. The 80s architectures, while they promised grand benefits, were never adopted by the government's space program. The final architectures of the SEI era were actually steps backward because their focus changed from development of lunar resources to intangibles, science and trying to do Mars while making the Moon a sideshow. It is clear from a cursory view of the federal budget that funding is available for things that are perceived to matter to the American people. A recent highway bill merely quibbled whether or not the number was between $218 billion and $375 billion. Just the difference between these two numbers could return us to the Moon. This is the connection that all of the previous studies and commissions missed.

What do we learn from all of the failures? Where did all of the previous work go wrong? The work of the space advocates from the 70s actually gives a clue to the core issue. The advocates, the L5 society, the National Space Institute and its merged successor, the National Space Society, as well as groups like the Space Studies Institute of O'Neill and its child the Space Frontier Foundation, have the vision. It is a simple statement on the Space Frontier Foundation's website:

Dedicated to Opening the Space Frontier for All Humanity

How does one do that? Can the government do this all by itself? The answer is no and the evidence is all of the government's failed efforts over the years. The government and NASA must come to grips with these truths or this new initiative will go the way of all of the previous ones. Space Frontier Foundation's vision statement is all encompassing in its scope, but it is incomplete as a guide to implementation. Several of the architectures in the 80s agreed with the Foundation's vision statement and they failed as well. So what else is needed? The mission statement at the beginning of this book is a start. Here it is repeated:

> *The exploration and development of space, including a return to the Moon and on to Mars, must bring concrete benefit to the people of the United States and the world, to improve our daily lives.*

This is what the first 8 chapters dealt with by discussing why transcending the limits to growth was important and how platinum mined from the Moon and used to enable the hydrogen economy would establish the vision and enable the mission. Today anyone who watches the news can understand how oil and the troubles associated with it are a threat to world stability. Also, with the growth of the economies of the world's most populous nations, China and India, will put further strains on the world's resources and political stability. We can blend these problems with the great joy of exploration, discovery, development, and settlement of the Moon and beyond.

Lunar Mining Value to the Terrestrial Economy

To meet the U.S. Department of Energy's goal of a total transition to the hydrogen economy by the early 2040s, a huge amount of platinum will have to be mined; more than five times the amount mined on a yearly basis today (refer to chapter 7 figure 7.4 and 7.5). According to calculations that I made based upon input from the study by the British government, this amount of PGM mining on the Earth will require gigawatts of electrical power, terajoules of natural gas and would generate hundreds of millions of tons of waste per year, much of it toxic. Developing the resources on the Moon allows us to shift a lot of this burden off the planet. Therefore it is important to understand exactly how much is there from the asteroids and what it is worth. Table 13.1 shows the relative concentrations of PGMs from various asteroids:

Metal	LL Chondrite	90th % Ni/Fe	98th % Ni/Fe
Germanium	1020	70	35
Gallium*	N/A	N/A	87
Platinum	30.9	28.8	63.8
Ruthenium	22.2	20.7	45.9
Rhodium	4.2	3.9	8.6
Palladium	17.5	2.6	1.2
Osmium	15.2	14.1	31.3
Iridium	15.0	14.0	31.0
Gold	4.4	0.7	0.6

Table 13.1: PGM and Other Valuable Metals Concentrations in Grams Per Ton[ii]

(*Gallium concentration from J. Lewis, Resources of Near Earth Space, p. 534)

One trait shown here that was not brought out in chapter 8 is that there are PGMs in all of the different classes of asteroids that exist in near Earth space and in their resulting impacts on the Moon. The LL Chondrite shown in column 2 in Table 13.1 is a type of asteroid that is as common as metal asteroids, but has a far lower overall concentration of metal. Therefore, it will be much more difficult to find their impacts on the Moon and so, in order to establish a conservative case, I am not considering them here. Also, their material strength is far less than a hunk of nickel/iron and so, are much more likely to have been completely destroyed in the impact and their materials spread into diffuse particles, thus rendering the resulting resources more difficult to economically extract. I am also not using the more optimistic concentrations for platinum. I will use twenty grams per ton average concentration although the potential for much richer resources is high.

An average concentration of 20 grams per ton of PGMs multiplied by 1 billion grams per year (1,000,000 kilos or 1000 metric tons [220,000 lbs or 35.2 million ounces]returned to the Earth) would require processing fifty million tons of nickel/iron asteroid per year from the lunar regolith and fragments that we find as a result of remote sensing. The total estimated inventory on the Moon of 140-590 billion tons (as estimated in chapter 8) is approximately a 3,000-year supply. At a value of $295 dollars per ounce delivered to the Earth, this works out to $10.4 billion dollars per year in revenue.

That is a lot of processed metal to get ~$10.4 billion dollars per year. However, this is no more processing than what will be required to mine an equivalent amount on the Earth in the near future. In this conservative case, platinum alone might not meet an economic test by itself. However, the processing of 50 million tons a year of nickel/iron meteorite would produce large amounts of palladium, osmium, iridium, ruthenium, gold, gallium, germanium, chromium, zinc, and other residual elements.

The total value per year of all of these resources could easily double the $10.4 billion dollar amount. The more rare PGMs, osmium, iridium, ruthenium and palladium, all have tremendous uses in industry and in fuel cells. There are no lack of applications of terrestrial PGMs, only supply. Based upon known meteorites, we can, with fair confidence, make a total estimate of extremely valuable metals using a very conservative discounted value of the PGMs and other metals. Table 13.2 shows the current price of PGMs and other resources, and the estimated discounted price for lunar derived PGMs and other resources because of increased production:

Resource	Price 2004 (Kg)	Value Per 1 Million Kg Today's mkt	Lunar Discounted	Value Per 1 Million Kg	Percent Global Supply/yr
Platinum	$20,811	$20.81 billion	$10,405 0.3 x	$10.4 billion	5X
Palladium	$10,226	$10.23 billion	$2556 0.25 x	$2.56 billion	5X
Ruthenium	$1961	$1.96 billion	$490 0.25 x	$490 million	12X
Rhodium	$26,464	$26.46 billion	$6616 0.25 x	$6.62 billion	15 X
Gold	$14,109	$14. 11 billion	$3527 0.25 x	$3.53 billion	.19 X
Iridium	N/A		N/A		
Osmium	N/A		N/A		
Germanium	$470	$470 million	$117.5	$117 million	22 X
Gallium	$530	$530 million	$132	$132 million	18 X
Total		**$74.57 billion**		**$23.85 billion**	

Table 13.2: USGS Metal Prices, Lunar Discount and Total Export Value

The demand curve for platinum will always be high, so I only gave a discount value of 0.3 for it. All of the current prices are from the United States Geological Survey.[iii] There is a class of fuel cell catalyst, which is made from a cocktail of platinum, ruthenium, osmium and iridium, that has four times the efficiency of a simple platinum catalyst. [iv] If PGMs were plentiful, then it could drive the demand curve for this advanced fuel cell and the materials that make it work. This could improve the macroeconomics of fuel cells well beyond current expectations. A mass-produced, affordable fuel cell, with four times the efficiency of a platinum-only cell, would save billions of dollars in fuel costs. Another remarkable property of this fuel cell is that the catalyst is good enough to run directly off of hydrocarbons as well as hydrogen. While this is speculative, it is based upon real resources known from meteorites found on the Earth, possibly on the Moon, and is indicative of what is possible when we are no longer limited in our resource options.

The dollar values for the discount rates on PGMs are conservative. We do know that the demand is going to increase dramatically as we go to the mass adoption of fuel cells and the hydrogen economy over the next 50 years. This is a reasonable assumption, even based on the conservative estimates of the oil reserves. Indeed, it is my contention that these lunar resources and their discounted value could actually be the bridge that makes the hydrogen economy practical. This is the reason to use the discounted value. While it would certainly make the economic argument to keep PGM prices high, the high prices today are what form the barrier to the mass adoption of fuel cells. Optimists in the past have used unrealistic market prices when flooding the market with large additions in supply but even with the large increase in supply, the demand is going to pace that increase.

The amount of materials that we are talking about returning to the Earth is not that high. A billion grams of platinum is only a thousand metric tons per year or 220,000 pounds. This is no more than what one Space Shuttle weighs. In developing a cislunar economy, much more mass will be moved per year in fuel, so transportation is not that big of an issue after the infrastructure is in place. However, in designing that infrastructure, we have to ask if PGMs are enough to make a cislunar economy work. It could go either way, although a direct comparison with the infrastructure and end to end cost of mining platinum on the Earth (including support infrastructure in power, pollution and loss of land use) would make platinum look much more favorable. Platinum mining will be the anchor of the lunar economy. It will facilitate other applications that will allow a profitable commercial lunar base with a positive balance of payments to the Earth.

One thing to stress here is that, with the discount value that has been applied in Table 13.2 above, the lunar platinum market is effectively subsidizing the implementation of the hydrogen economy. Today, the average raw cost of platinum production in South Africa's Bushveld complex, one of the richer ore bodies, is ~$196 per ounce. This price does not include paying back the investment for the equipment needed to mine and produce the ore.

In the above analysis, I am postulating a price of $295 for lunar platinum, which implies a production cost of no more than the terrestrial competition. If we use a number that is half the current market price of $800 per ounce, then the revenue per year would rise to $14.1 billion per year, which would probably be profitable by itself. As of June 10, 2004, the daily spot price is above $800 per ounce. This number is high in recent experience and will probably fall but even the $400 per ounce number is well below the last ten year historical average. Therefore, at a price of $295 per ounce, which is a seventy percent discount over today's market price and also well below historical prices, a fuel cell "engine" for an automobile would be much more competitive compared to the cost of an internal combustion engine. If you look at the much greater fuel efficiency of fuel cells, the balance starts to decisively tip in their direction. Also, greater terrestrial demand today is well outpacing the global supply and many of the primary producers have had problems increasing their output.

In looking at this in a "gestalt," "holistic", or "grand scheme" context, there is a bright future for lunar PGMs. Nevertheless, there are some who are prominent in the planetary science community that dispute that any lunar nickel/iron fragments exist other than the very diffuse particles already discovered in the regolith. The evidence presented here seems to defy this conclusion, but in the end, this is a scientific hypothesis, one easily tested with a lunar orbiting spacecraft backed up by a lander to assess the local resources found from orbit. This is the underlying merit of this book's thesis. It is supported by existing evidence and it is testable with instruments of general value to lunar science. Testing hypotheses is the nature of science. It has been less than 40 years since geologists were laughed out of rooms for postulating that the Sudbury mining district, the source of hundreds of billions of dollars worth of nickel, copper and PGMs, was the result of an asteroid impact. Today, nickel and PGM discoveries are being made using impact geology as a guide. The following is from a recent BBC article on the subject:

> *"On average I would say that one quarter of the known impact structures on the Earth have some sort of deposit associated with them," Canada's Natural Resources Department chief scientist Richard Grieve told BBC World Service's Science In Action programme.*[v]

The question that has to be asked of the planetary scientists is: If this process has clearly happened on the Earth where the average impact velocity is higher, then why has it not happened on the Moon? My answer is that it has and the first mission in any new initiative should be a new Lunar Prospector with at least one instrument devoted to this task.

A Lunar Architecture for Cislunar Economic Development

Departure from Previous Justifications for Lunar Exploration and Development

At this point it is necessary for me to begin the process of laying out my own architecture for the return to the Moon before going onto the other applications that positively influence the economics of a lunar base and cislunar economy. In developing an architecture, first you must articulate your objectives. In this process, I am going to take issue with many of the previous efforts. Their economics and ability to make money for Earth were questionable. Rarely before have lunar development advocates postulated activities that could make money without a scale of investment that only a government could possibly make. This is the reason that none of these have succeeded in the past and so this is to be avoided if at all possible.

In political discussions, advocates of the Kyoto Accord and the hydrogen economy postulate that hundreds of billions to trillions of dollars will have to be spent on carbon dioxide pollution mitigation and the transition away from the oil economy. This is to solve the assumed problem of global warming. The hydrogen economy has been touted as the solution to the problem. The hydrogen economy has also been sold as the way to transition away from the oil economy as that resource is depleted. Global warming, if it is from carbon dioxide pollution, and resource depletion are the two largest global problems that confront civilization today. Therefore, my postulate is that the development of lunar PGM resources is the first step to solve the global warming problem by jumpstarting the hydrogen economy. This brings space development front and center in today's largest issues, but in doing so, sweeps away many of the prior justifications for going back to the Moon, or relegates them to secondary or tertiary status.

An ancillary benefit to the production of this amount of PGMs is the 50 million tons of "waste" cobalt, nickel, and iron. A conservative estimate is that 50,000 tons of cobalt will be produced per year. As for nickel, this will vary widely, as the data from samples of meteorites on the Earth indicate, but there will be a minimum of 5 million tons and the possibility of as much as 30 million tons of nickel per year! The rest is iron, amounting to between 20 to 45 million tons. To compare this with terrestrial production, according to the U.S. Geological Survey, in 2003, global nickel production was approximately 1.4 million tons.[vi] The number for iron is about 634 million tons. While at present it is impractical to export this mass of metal to the Earth, it is intriguing to think about the new technologies that can make it possible within the next few decades. The time may, and probably should, come in the future that all terrestrial production of primary metals can be shifted off planet. This would be an incredible boon for the environment as well as the ultimate justification for developing the resources of the solar system. The Moon is just the beginning for such developments since all of these resources are derived from asteroid impacts. With asteroids, such as 3554 Amun, with over 3 billion tons of metals and Cleopatra with a billion times more than Amun, we are in no danger of running out of resources in the next few millennia.

PGM mining on the Moon will be the enabler for a vast lunar economy, especially if power rich methods are used for separating the PGMs, cobalt, nickel and iron from each other. The nickel/iron/cobalt mixture could be poured into moulds, formed into beams and used to build structures, roads, railroads and large enclosed spaces. With these metals, we can build a railroad that could encircle the Moon, providing the beginnings of a lunar transportation network. Very tall skyscrapers could be built on the Moon from materials similar to those that we use to build them on the Earth. These structures will never rust in the vacuum of space and a covering of three feet of iron is the equivalent in radiation protection of the Earth's atmosphere. Vast quantities of high strength metals can be lofted from the Moon's surface and used to build an L1 space station of enormous proportions and eventually something the size of an O'Neill colony.

This industrialization will obviously need more than a few employees to execute. This is the final link in the chain that we have been building. Space is for the people. For people to support this, they have to feel that, not only do they have a part, they have the possibility of participating, not just watching it on TV or seeing it on the Internet. An extensive lunar base and activities at multiple lunar locations will require people, a fair number of people. Within 30 years thousands of people could be living and working on the Moon and even beyond. Within the next 50 years we could have many times this amount living and working on the Moon, Mars, Asteroids, and the future reaches of the solar system. To paraphrase Bobby Kennedy: People look at the Moon and ask why? I look at the Moon and its resources and say why not!

Examples of Prior Justifications

One of the biggest commercial justifications that has been talked about is solar power beaming from the Moon to the surface of the Earth. Dr. David Criswell of the University of Houston has been a proponent of this for 30 years. The idea is to manufacture silicon solar cells out of lunar derived silicon, assemble them into a large-scale power generation system and then beam this power back to the Earth either via laser beams or radio waves. While the idea of building large power systems on the Moon out of native materials is a good one, the power would be much better used locally on the Moon for processing the nickel/iron and extracting the platinum and other valuable metals. In a way, this is providing power to the Earth but just not in the way Criswell originally envisioned.

Referring back to chapter seven where we looked at the terrestrial resource demands for refining platinum, we find that for a billion grams, or one million kilograms, the requirements for power, natural gas and the resulting waste production is:

6.4 Million Tons of CO_2
23,760 Gw/hr of Electrical Power
10.45 terajoules of Natural Gas
250 Million Tons of Waste Rock

These are all yearly numbers. If a five gigawatt solar power generation station was built on the Moon for transmission, no more than one gigawatt would be transferred to the power grid on the Earth. The amount of power needed for refining one million kilos of platinum on the earth is about 2.7 gigawatts running all year. Even if you needed the whole five gigawatts from the lunar power station for melting the nickel/iron and separating out the PGMs on the Moon, you would be displacing all of the above resource requirements and pollution on the Earth! Additionally, the pollution and waste on the Earth is much worse than this in its entirety due to the harsh chemicals that are used to refine platinum and the rich farmland of southern Africa that is sacrificed to feed the needs of the developed world. Using the power on the Moon would also eliminate the need to build vast arrays of receivers on the Earth, also an expensive proposition. Also, there would be none of the environmental concerns associated with beaming large amounts of electrical power through the atmosphere.

Using the power on the Moon to create resources for the Earth will save the 2.7 gigawatts of power on the Earth per year, conserve natural gas as well as eliminate millions of tons of CO_2 pollution. The "power" is "transmitted" as the refined platinum to the Earth for our use here. This proxy transfer of power eliminates the need for lunar based power beaming. It eliminates the infrastructure cost for developing and emplacing the transmitters or the lasers as well. In effect we are transmitting 2.7 gigawatts of power continuously throughout the year in the form of the platinum.

Helium-3

Dr. Harrison Schmitt, of Apollo 17 fame and a leading voice at the Fusion Technology Institute at the University of Wisconsin, has been a long time advocate of the mining of helium-3 on the Moon for use in clean burning, radiation byproduct free, nuclear fusion energy.[vii] The Fusion Technology Institute understands and shares the vision that I present here about how resources from the Moon will help solve our most difficult problems on the Earth. This is illustrated in this excerpt from their vision statement on their website:

> There is no doubt that one of the most difficult problems that a peaceful world will face in the 21st century will be to secure an adequate, safe, clean, and economical source of energy. Existence of lunar helium-3, to be used as fuel for fusion reactors, is well documented; verified from numerous Apollo and Luna mission samples, current analyses indicate that there are at least 1 million tonnes embedded in the lunar surface. The helium-3 would be used as fuel for fusion reactors.

I explored this concept earlier in chapter 5 and I absolutely agree with them on the usefulness and value of Helium 3 for nuclear fusion reactors. However, there is a critical flaw in their plans to do this. The environmentalists would go nuts over the complete disfigurement of the Moon to extract that 1 million tons of Helium 3 from the regolith. That would make the picture in Figure 11.10 look tame by comparison. The fact is that we will have to mine some regolith in the early days of a lunar base simply to provide oxygen, titanium, aluminum, silicon and other resources needed to build a lunar civilization. However, since these same resources are in all of the rocks of the Moon, all that will ever be needed can be obtained from deep mining rather than having to strip mine the whole surface of the Moon. The same cannot be said for helium-3 that is deposited by solar wind on the Moon's surface. We need to look beyond the Moon for this tremendous energy source. This is what I talked about in Chapter 5 concerning Dr. John Lewis's proposition to obtain unlimited quantities of helium-3 from Uranus and Neptune.

The Moon does have a role in obtaining this resource. President Bush actually talked about building the spaceships on the Moon that will go to Mars in his January 14, 2004 speech on the new exploration vision. It is just as tenable to postulate that the spaceships that will go to Uranus and Neptune to obtain the much more plentiful helium-3 will be built on the Moon. This would probably be much cheaper than the mass disruption and cooking of the lunar regolith to obtain the extremely rare helium-3. This can be just another application for the copious "waste" products derived from roasting the regolith in vacuum induction furnaces and other methods to obtain oxygen. Table 13.3 gives the percentage composition of both lunar Mare and

Highland regolith:

Oxide	Highland Regolith	Mare Regolith
SiO$_2$ (Silicon Dioxide)	44.8	38.9
TiO$_2$ (Titanium Dioxide)	0.5	8.3
Al$_2$O$_3$ (Aluminum Oxide)	28.1	16.1
FeO (Iron Oxide)	4.2	16.2
MgO (Magnesium Oxide)	5.5	8.3
CaO (Calcium Oxide)	15.7	11.5
K$_2$O (Potassium Oxide)	0.8	0.2
Na$_2$O (Sodium Oxide)	0.4	0.6

Table 13.3: Percentage Molecular Composition of Lunar Regolith

These resources are available for use by a lunar base, settlement or civilization. There are many possible methods of obtaining these resources that I will talk about later, but here, all that is necessary to know is that metals, such as Aluminum, are very common on the Moon, as are titanium and iron that are not of meteoritic origin. For example, for every ton of oxygen liberated from FeO, 3.25 tons of iron is made. If you use Al$_2$O$_3$ as your feedstock, you get 1.08 tons of Aluminum per ton of oxygen. For TiO2, the result is 1.375 tons per ton of oxygen. If Silicon dioxide or SiO$_2$ is used, then you get 1750 pounds of silicon per ton of oxygen. These are most of the metals that we use in building launch vehicles on the Earth and these could be combined with the waste metals from the PGM mining for shielding.

Spaceships built on the Moon would be a logical export product. This would be to build up a robust infrastructure for cislunar transportation activities. This would also cut the demand for ground based launches over what would otherwise be necessary. The key to lowering the cost of the total logistical chain between the Earth and the Moon is to maximize the use of lunar resources. These spaceships would have the benefit of having whatever shape makes the most sense for the application. Such spacecraft could be built in an optimal fashion for trips to Mars, the asteroids, or the outer planets. If spaceships are manufactured on the Moon, then the Earth becomes an exporter of high tech goods to the Moon such as computers, electronics for engine controls and other high tech items that it is not cost effective to make on the Moon.

The vehicles built for Helium-3 mining would be incredibly advanced in comparison with what we build today and would acquire Helium-3 with far less cost than strip mining the whole surface of the Moon. As we mentioned before, if 4,000 tons per year were shipped back from Uranus for use in terrestrial reactors, it would still take 4 **billion** years to exhaust the supply from Uranus, versus 250 years from the Moon's supply. This mining of Helium-3 on Uranus would leave no scars either since it is separated from that gas giant's atmosphere.

So Dr. Schmitt and the Fusion Technology Institute are right. Helium-3 is ultimately the way that we should go in developing nuclear fusion. We could even sell reactors to North Korea without any worries about them being converted for nuclear weapons production. However, lunar derived Helium-3 should only be sparingly obtained from other regolith processing activities and never in the quantities that they seek today.

Water

There has been much speculation about the existence of water in the permanently shadowed regions of the Moon. Some concrete evidence of this was obtained in both the Clementine and Lunar Prospector missions. Estimates in the scientific community range from one million to a billion tons. This water, if found, would bring major advantages to the development of the lunar economy. The water on the Moon is concentrated at the lunar poles where there are regions that have never seen the Sun and are, therefore, very cold. The best scientific evidence is that the temperatures there are no more than 50 to 70 degrees above absolute zero. Figure 13.1 is a graphic showing the expected areas where elevated hydrogen levels have been detected:

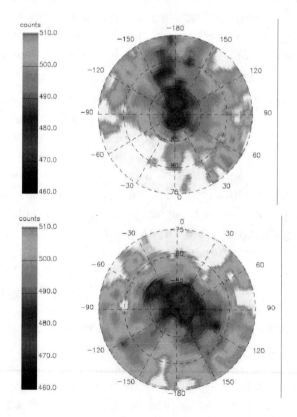

Figure 13a, b: Lunar North Polar and South Polar Hydrogen Concentrations

The numbers are somewhat misleading because what they are measuring are reduced counts of slow neutrons emanating from the surface of the Moon. These neutrons are generated by the impact of high-energy cosmic rays and the reduction comes from some of them being absorbed preferentially by hydrogen. This hydrogen is generally assumed to be bound with oxygen in the form of ices. These ices are located in these regions because comets and asteroids that were water and hydrocarbon rich have impacted on the Moon. The water would have vaporized and at least a very small fraction would have migrated to the poles as part of a thermodynamic process.

If this water truly exists, which all evidence seems to point to, it will be a great enabler for lunar development. Hydrogen will not have to be taken from the Earth to the Moon in the early days to mate with the oxygen generated from the known sources in the regolith. If the higher estimates of water exist on the Moon, it will be invaluable for transportation, lunar agriculture and for human consumption. However, it should not be widely used off of the Moon itself. It is far more valuable for use in developing the lunar economy over the mid-term. It will be a valuable resource in the early days of development for fuel for spaceships, but for the long-term, the most plentiful and valuable water resources will come from extinct comets whose orbits are similar to those of near Earth objects.

When you look up into the sky and see the beautiful tail coming off of a comet, most of that tail is water. As comets "age," they tend to develop hard-caked surfaces that do not allow the water to escape. This type of asteroid is rich in water and other volatiles, possibly several billion tons of it. The same spacecraft factory that would be built on the Moon for extracting Helium-3 from the atmosphere of Uranus would build dozens of both manned and unmanned spacecraft for exploring and using these inner solar system resources. Current estimates are that there are tens of thousands of these objects between the orbit of Venus and the main asteroid belt. Eventually, the Moon would become a net importer of water to feed its growing civilization. Also, there is tantalizing evidence that Phobos, the larger of the two moons of Mars, is a type of asteroid called a Carbonaceous Chondrite that could be mined for water as well for use for propellant near Mars.

One thing that the water, or the much more plentiful oxygen, from the Moon cannot do by itself is generate a lunar economy. Lunar water and oxygen only have value within the context of a vibrant cislunar economy. That economy has to have some other engine to run it. The water and the oxygen, while very important, play supporting roles that help to control the costs of the lunar infrastructure. In some of the architectures put forth in the 80s, lunar oxygen and by inference, water, would be used to enable low cost space tugs between the LEO and GEO orbit and LEO and the Moon. Unfortunately for the lunar architectures, the communications satellite industry was profitable without the added benefit of these lunar resources. Today, the same thing cannot be said for the hydrogen economy. Even with the heavily discounted price per kilogram proposed here, a lunar base can be profitable, if the infrastructure is set up in a cost effective manner. I will postulate one possible infrastructure in the next chapter. Lunar water and lunar oxygen will help lower that infrastructure cost and that is their true value.

Other Lunar Industrial Justifications

In the discussions of architectures and the exploration of ideas for lunar economic development, there have been certain constants that have appeared in the literature. These were dealt with in the previous section concerning lunar power beaming, helium-3, lunar water and oxygen production. However, in the early days of the Apollo program, there were other very interesting justifications offered for lunar economic development. A whole book could easily be written on the various ideas previously discussed as well as the lesser-known ideas, but one book caught my attention as a brilliant exposition on the possibilities of manufacturing on the lunar surface. This book was, "The Case for Going to the Moon," by Mr. Neil Ruzic. This book made the case for both research and manufacturing on the Moon by taking advantage of the ability to use the extremely hard vacuum to create stable temperatures.

On the Earth vacuum manufacturing is a multi-billion dollar industry. Many metal alloys, especially those of importance to the aerospace industry, are formed in vacuum chambers on the Earth. Turbine blades, made from exotic alloys, would not form properly in a normal atmosphere due to the reaction with oxygen, nitrogen and trace gases due to the heat involved. Semiconductor manufacturing also utilizes hard vacuum. The emerging field of nanotechnology is highly dependent upon high quality vacuum for proper operation of nano-fabricated actuators, motors and other parts. Figure 13.2 shows a vacuum manufacturing installation on the Moon as envisioned by Ruzic:

Figure 13.2: Lunar Vacuum Manufacturing (Picture Courtesy Mrs. Neil Ruzic)

This book espoused the development of an invention that he later patented, called the cryostat, a device used today in vacuum manufacturing. This Cryostat, shown in Figure 13.2, would form the basis for mass manufacturing of all sorts of lunar products as introduced by Ruzic. Cryostats are now standard equipment on scientific satellites with demanding low temperature requirements. Ruzic foresaw the lunar cryostat 40 years ago for use in cooling infrared and other sensors that are necessary to detect very diffuse or faraway sources. The book also postulated that, with the use of cryostats and the ability to virtually dial in any temperature desired, superconducting magnets could be operated without any power input.[viii]

Other applications of superconducting vacuums that were espoused by Neil Ruzic in his book include thin films, optical components, pharmaceuticals and the possibility to push forward the frontiers of nuclear fusion. In the 1980s, Krafft Ehricke postulated that it would be easier to build a test helium-3 reactor on the Moon because it would require much less power. No power would be required to cool the superconducting magnets. Also, since a very high quality vacuum is required for nuclear fusion, the vacuum vessel would be eliminated, making servicing of the system far simpler and allowing a much greater frequency of operations testing.

While vacuum manufacturing profits are extremely speculative at this time, the applications are very interesting and could become a vital part of a future lunar economy. One thing that is certain is that it will be possible to make a large number of exotic alloys that are not possible on the Earth due to the cost and demand for high vacuum.

Lunar Science

One of the strongest and longest lasting justifications for returning to the Moon is related to science. There have been many papers and studies about the value of lunar-based telescopes for optical and infrared wavelengths. Radio telescopes on the lunar far side have also been proposed. The justification has always been the stability of the lunar surface, constant temperature and, for radio telescopes, the shielding provided by the Earth from radio interference. These scientific uses of the Moon have been, and continue to be primary justifications. The problem is that, as was stated by some of the lunar scientists, in order for these telescopes to work, industrial activity would be precluded. The activity of landing and taking off would deposit dust on optics, and radio traffic in the vicinity of the Moon by commercial interests would ruin the qualities that make the Moon desirable for lunar based radio science. If the development of the resources of the Moon is our goal then these scientific justifications no longer apply.

The James Webb Space Telescope (JWST) currently under development by NASA is a large infrared telescope that will orbit the Earth/Sun L2 libration point. This is just the start of a trend toward large telescopes in space. In the future these telescopes will continue to grow in size and complexity. Figure 13.3 illustrates NASA's latest roadmap toward larger space telescopes:

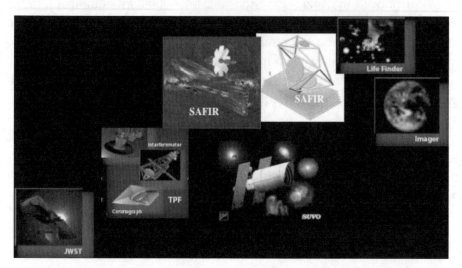

Figure 13.3: NASA Roadmap to Larger Space Telescopes

It seems that NASA has bypassed the Moon in their future plans for larger telescopes and, ironically, the technology that is allowing them to do this is based on the lunar cryostat invented by Ruzic. Also a factor, is that, as the size of telescopes scale, such as one that would be required to actually image Earth sized planets around other stars, the gravity of the Moon is just as much an impediment as it is on the Earth itself. As a telescope gets larger the tolerances do not get larger as well. An optical surface at visible light wavelengths has to be smooth to within 1/8 of a wavelength no matter if it is 5 inches or 50 feet in diameter. The largest telescope we have ever built on the Earth is 8.3 meters in diameter. To actually image Earth sized planets around starts a hundred light years away (which is NASA's goal) will require a telescope mirror 50 meters in diameter! Also, guidance technology being developed by the JWST program makes the requirement for the stability of a platform on the Moon obsolete.

As far as radio science is concerned, the Moon today is not radio quiet because spacecraft from all over the solar system transmit back to the Earth, which also impinges on the Moon. This will get worse as NASA's Mars exploration program moves into higher gear, with multiple spacecraft in orbit around Mars, spacecraft in orbit around Saturn and multiple spacecraft such as Voyager still transmitting from beyond the edge of the solar system. Radio science telescopes will be required to move farther and farther toward the edge of the solar system and away from the Moon and the inner solar system.

Other science on the Moon besides telescopes will be dramatically aided by the development of the Moon. Any government funded exploration program will, of necessity, be limited by the amount of funding available. If the past 32 years, since the end of the Apollo program, is any guide, those funds will be either limited or nonexistent if the Moon is made the preserve of scientists. While no one, including me, proposes, postulates or predicts that massive scarring of the lunar surface will ever occur, there will be some activity. None of this activity will sully the surface of the Moon to the extent that it would preclude scientific study of the regolith for the history of the output of the Sun. Also, the history of the Moon's structural evolution will be enhanced rather than precluded by lunar development.

The Moon is far too valuable to the future of mankind to make it an extraterrestrial preserve for science, and there is no activity that can take place on the Moon that precludes scientific investigation. Telescope technology is leaving the lunar telescope builders behind.

The Moon for the People

A friend of mine recently wrote on an Internet bulletin board that he would be proud to clean the bathrooms at any lunar base. This is a common sentiment, shared by many people. They would do whatever it takes to support getting to the Moon if they had any chance of getting there themselves. While this is a sincere sentiment, I am quite sure that there will be many more exciting jobs on the Moon.

During the Apollo era, over 500,000 people worked on the task of getting mankind to the Moon and back safely. I cannot get out of my mind a meeting that I had last year with an old gentleman who was the night manager of a casino/hotel outside of Reno Nevada. It was very late and there were no rooms available. As we were talking, we discovered that we had a mutual interest in space. It turns out that he was a test engineer for the pyrotechnics that separated the Apollo Lunar Excursion Module ascent and descent stages at the critical point on the lunar surface when the ascent module took off. He told me that his only job was to verify the reliability, test record and all of the quality control associated with every one of those 121 pyrotechnic devices. He personally inspected and verified every single one of them in his job at NASA before the LEM was loaded on the Saturn V in Florida. He knew that if he did not do his job right, that Neil Armstrong and Buzz Aldrin would die on the Moon, and that the entire Apollo program at that moment depended upon him and his competence. You could see his old and fading eyes blaze up in justifiable pride at the sure knowledge that, because he did his job, Neil and Buzz made it safely back from the lunar surface. Today it is no longer adequate to have participated in the way that that gentleman did. Today, we want to go into space ourselves. We want more than to watch the few and the brave go for us.

A few weeks ago, at the final public hearing of the Moon-to-Mars Commission, better known as the Aldridge commission, some remarkable testimony was heard. George Whitesides, Executive Director of the National Space Society, the successor to the National Space Institute founded by von Braun had some fitting words to say about people and the Moon.

...we must all evangelize an exploration society predicated on settlement. This

ultimately is the real cause of the exploration that we seek. To create a space-faring civilization, a civilization of vibrant communities living and working beyond Earth. This is the crucial link which ties together space exploration, private enterprise, and public participation. Settlement is the destination for exploration's efforts. Without it exploration is a dead end.

George is absolutely right in this call for settlement. Another speaker, Dr. Anthony Tether, head of the Defense Advanced Research Agency DARPA, echoed the betrayal that many of our generation felt at the end of the Apollo program:

...what NASA seemed to forget was that we all wanted to go. We all wanted to go and somehow we lost that. We all wanted to go and it was almost taken away. How many people believe that they can go? And until we get that excitement back, nothing much is going to change...We all wanted to go and we were forgotten about. Those who wanted to go were forgotten.[ix]

We all wanted to go; great sentiments from the head of the Defense Department's famous space research and development laboratory. This is the key for public support. However, public support is not enough to sustain and build a lunar civilization. It has to pay its way just as the United States in its early days, as a colony of Britain, had to pay its way. This is why the mining, the resources and the overall focus on the development of the solar system, beginning with the Moon, have to be the main reasons for going. Even more importantly, the development of the vast resources of the Moon and beyond is the way to truly transcend the limits to growth. The proponents of limits to growth strongly and accurately believe that the resources of the Earth are limited. However, the riches that await us on the Moon and in the rest of the solar system illustrate that the only real limit is our imagination.

In the next chapter, I will propose a lunar architecture that I believe will be cost effective and possible to implement with a minimum of government support. Private enterprise cannot yet shoulder the whole burden, but, just as the U.S. and state governments supported the birth of the steam ship, the railroads, the Panama Canal and our nation's current air and road transportation network, so can they support the development of our cislunar infrastructure. That argument I will save for the final chapter.

*   *   *   *   *

[i] L. Haskin, et al, Geochemical Assessment of Lunar Ore Formation, Resources of Near Earth Space, University of Arizona Press, Tucson, 1993, P. 26
[ii] B. Blair, *The Role of Near-Earth Asteroids in Long-Term Platinum Supply*, EB535 Metals Economics, Colorado School of Mines, May 5, 2000, P 7
[iii] http://minerals.usgs.gov/minerals/pubs/mcs/
[iv] E. *Reddington, et al, Combinatorial Electrochemistry: A Highly Parallel, Optical Screening Method for Discovery of Better Electrocatalysts*, Science, Volume 280, June 1998
[v] http://news.bbc.co.uk/2/hi/science/nature/3281611.stm
[vi] http://minerals.usgs.gov/minerals/pubs/commodity/nickel/index.html#mcs
[vii] http://fti.neep.wisc.edu/research/he3.html
[viii] N. Ruzic, The Case for Going to the Moon, G.P. Putnam Sons, New York, NY, 1965, P. 60-65
[ix] T. Tether, The President's Commission on Implementation of the United States Exploration Policy, Asia Society, 725 Park Avenue, New York, NY, May 3, 4, 2004, P. 34

Chapter 14:

A 21st Century Commercial Lunar Architecture

Rethinking the Moon

To people who have been associated with, or watched, NASA's futile efforts to return to the Moon for the past 32 years, the title of this chapter seems absurd. Space exploration is expensive, it brings no returns to justify the investment other than scientific knowledge and therefore, will always be the province of government, or so the conventional logic goes. The last forty-six years of mostly futile efforts of mankind to move out into the solar system can be traced to one source: the government. To paraphrase Ronald Reagan:

> *Government is not the solution to our return to the Moon problem, government is the problem.*

This is not because NASA is bad or evil. It is the nature of a government agency to be like NASA. NASA is also a victim of being a government agency. It is dependent upon the whims and fads that sweep the congress just like they do in the general society. As the conventional wisdom goes, "If the congress would just do multi-year appropriations NASA could plan." This argument has been made for 40 years and is no closer to happening now than it was then. Another argument goes, "if we just had leadership from congress or the president, then we could move on a vision for space." While this is true, it is also the problem. SEI, and the exploration initiative of the elder President Bush, died on January 20, 1993, when a new leader came in who did not share that particular leadership vision. There is every chance that an incoming president may not share the vision announced by George W. Bush in January of 2004. This is what NASA has always faced and always will.

NASA, as a federal agency, is bound to adhere to all sorts of government regulations that drive up cost and breed inefficiency. This is also true of the contracts with industry that they award. The Federal Acquisition Regulations (FARs) contain thousands of pages of rules and regulations that govern how contractors do their job, allocate subcontracts and even the skill levels of individual workers for the larger contracts. NASA also is the victim of frequent budget changes that today awards a billion dollar contract to a Boeing or Lockheed, and tomorrow, have it taken away when a the agency decides to go in a different direction. This happens even more often internally at NASA. No civil servant's career is complete without having many programs cancelled. I know of one program manager's mission that was on and off again for ten years before final approval. The recently launched Gravity Probe B mission was started in 1968! The combination of regulations and the whims of political fortune make it hard for even the most dedicated NASA civil servants to do their jobs.

In some ways, NASA is an American historical aberration. It was born at the dawn of the technocratic age, when those in government bought into the notion that the purpose of the federal government was *"to set goals for the nation and devise methods for their achievement under state direction."*[i] These political thinkers believed that the *"power of the state was not corrupting, but a tool to be used for good."* In the arena of space, NASA was the tool of the state to execute great deeds in winning the marketing campaign of the cold war. Space played into the mindset of the influential leaders of that era who were sold on the idea that the prestige of American exploits to reach the Moon would sway the masses in Asia and Africa to believe in the superiority of our American system. MacDougall makes the following observation concerning prestige in his book on the political history of the space age:

> *Prestige as we now understand it is a recent usage. The word originated in the Latin praestigium, an illusion or delusion, usually rendered in the plural to denote "juggler's tricks." In French and then English the word meant deceit: a "prestigious man" was a fraud.*[ii]

NASA was at the point of the spear in the "prestige" wars of the 1960s. After it served its purpose, it was discarded for the next set of pretty images, that of the environmental movement and the "war on poverty," all set to the same tune of "we are from the government and we are here to help you." NASA has never recovered from the blow. However, after the layoffs in aerospace in the 1970s as the Apollo dream died, a miracle occurred that sidestepped the image and changed the world in a little place that would soon be called Silicon Valley.

The Silicon Valley Way?

In the early 1970s, hundreds of thousands of engineers, scientists, technicians, draft persons, assemblers, testers and others who made up the heart and soul of the space program were laid off, a great number of them in California. At the time that space was dying as a mass industry, a revolution was brewing in both Northern and Southern California. It began as a seedling in Silicon Valley, when engineers named Noyce and Moore left defense and space contractor, Fairchild, to found a small company named Intel. It began growing when an ex-United States Air Force officer, named Ed Roberts, blew his company's last dollars (they had made kits and test equipment for the nuclear program at Sandia labs, but were dying in the downturn) and started selling a kit computer, based on the Intel 8080, named the Altair. The word's "BASIC" as software and "microcomputer," as hardware would begin being used together when a pair of school friends wrote to Altair on their fictitious high school company letterhead, Traft-o-Data, saying that they had a BASIC compiler for the new kit computer. That company was later renamed Microcomputer-Software, or Microsoft.

The revolution gained soldiers for the army that would change the world in a garage in Silicon Valley where the Apple I was invented by a kid named Steve Wozniak and his friend Steve Jobs. This computer helped begin the second phase of the revolution. The huge success of this computer awoke the entrepreneurial spirit in the director of the famous Hughes Research Laboratory, Dr. Robert S. Harp, his wife Lore, and Carol and Bob Ely. They built the first high-resolution graphics board that would pave the way for Chinese characters to be generated on a microcomputer. They jumped into the microcomputer fray with the computer near and dear to my own heart, the Vector I. This was the computer, compatible with an Altair, that ran Microsoft BASIC programs made for business uses. This computer was the first microcomputer to inhabit the offices of lawyers and doctors.

All of these companies were founded between 1970 and the end of 1976! Other companies followed. Cromemco, NorthStar, Pertec Computer and Digital Research were all familiar names in the late seventies as the microcomputer world spun faster and faster. Peripheral companies, such as Micropolis, Tandon, Seagate and Quantum, were founded to provide disk drives to feed the exploding demand for storage. Companies that built video terminals, keyboards and cables all began growing. Half of the staff of these companies were people who had only recently been building Moon rockets, missiles and space stations. The other half were the younger brothers and sisters of the hippy generation who grew up on Apollo, Star Trek, ham radio and homebrew computers. There is a famous picture of the staff of Microsoft in 1978 that says it all about how we were back then.

Figure 14.1: Microsoft Corporation 1978

There is a famous caption that goes along with this picture that illustrates, even beyond this picture, why that period was a revolution. The Caption was, "Would you invest in this Company?" This simply illustrated the vast gulf between the technocratic, government directed aerospace world and the world of the microcomputer industry of the late 70s and early 80s.

If you had invested a thousand dollars in Microsoft in 1978, which would have been
8 years before their Initial Public Offering, you would be worth more than a hundred
million dollars today. Since 1985, Boeing has had four stock splits. Microsoft has had
nine stock splits. Seven of the nine stock splits at Microsoft have been two-to-one
splits. At Boeing, only one of the four stock splits was two-to-one. The people in that
picture above have a cumulative net worth close to 100 **billion** dollars. Look at the
group above and think about this for a moment. Figure 14.2 illustrates the
comparative difference in return on shareholder value between Microsoft and Boeing,
the world's leading aerospace contractor to the government.

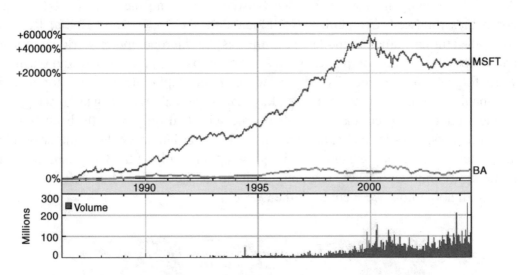

Figure 14.2: Comparative Stock Price Between Microsoft and Boeing 1986-2004
(Microsoft at top (MSFT) and Boeing at bottom (BA))

With all of the new microcomputer companies came jobs, lots of jobs. When I moved
to California in 1980, demand was brisk for any human who knew which end of the
integrated circuit was up. In 1981, when I went to work at Vector Graphic Inc, in
Westlake Village, California, as a test technician, I worked 80 hour weeks along with
all my compatriots. During the recession of 1982-83, the size of our company more
than doubled. Compaq was founded during this time along with many of the PC clone
companies that swiftly became billion dollar companies. It was an amazing time to be
working in an industry. The growth and excitement was everywhere! While the rust
belt was emptying of jobs and people, we were partying at Comdex in Las Vegas.

I was at "Vector," as we called it, for a grand total of three days before I received stock
options. I had never heard of stock options before! Lore Harp, the company
president, met each of us at the front door of the company and gave us Vector T-shirts
that read, "Computers for the Betterment of Society." She, her husband Dr. Bob, and
the other owners, took us all to the beach at Malibu in several buses for a wine and
cheese party! We were a dot com company long before the dot com boom even

existed. We also knew that we, along with everyone else in the microcomputer industry, were changing the world, and as long as I live, I will cherish the memories of that time and what we did.

Some companies grow to become Seagates, Microsofts and Apples, while others, like Altair, Micropolis, Digital Research and Vector Graphic, become footnotes in history. That is the way that the world works. That is the winnowing that takes place. The race is not always to the best product either, the PC being a prime example.

The corporate behemoths saw what was happening and also entered the fray, as IBM did in August of 1981. However, that entry into the industry did as much to change IBM as IBM did to change the microcomputer industry. The hippies and the suits together brought legitimacy to the business of computers, and together, through many trials, helped build an industry with global revenue larger than the gross national product of all but a few nations. This revenue is also several times larger than the total global aerospace market and Microsoft has more cash in the bank than the next three years of NASA's total budget.

How did Silicon Valley beat the aerospace world? Quite simply, it was substance over image. When the federal government was requiring a certain number of PhDs on a government contract in order to fill some nebulous requirement to build a spacecraft, in silicon valley, non degreed technicians, like myself, were taken out of manufacturing and trained to be hardware and software engineers. At Vector Graphic Inc. in 1981, at the height of the company, we had seven engineers. Only one had a degree. Other companies had similar stories. If you had a brain and were willing to work hard, you not only could make money, you could make products that changed the world.

That was the legacy of Silicon Valley for years and years and still makes up a part of the culture. There are more degreed engineers now and the market is much more mature, but there is still a culture that rewards innovation. The dot com boom was just the second phase of what we started in the early 80s. New names such as Musk, Forina and Bezos dominate the industry and have built their fortunes. While Noyce and Moore may be retired now, we still have Gates and Jobs as forward looking innovators, and Bill Gates' friend, Paul Allen, is funding another maverick, Burt Rutan, to do what no other private company has ever done…fly into space! …and thus the circle closes and a possible solution emerges.

It is not at Boeing or Lockheed that the future will be written. It is in companies like Scaled Composites (funded by the guy in the lower right corner of the picture in figure 14.1), SpaceX.com (funded by the dot com boom's Elon Musk), Armadillo Aerospace (funded by John Carmack and the revenues from Quake!) and other entrants not as well funded or known today.

NASA cannot open the space frontier alone and the evidence is that it cannot do it at all unless dramatic changes occur. NASA has a role model for its future in its Silicon Valley analog, IBM. IBM moved from being the epitome of the stodgy, corporate world, to a nimble intellectual property powerhouse whose products form the foundation that underpin much of the computer world today. With the release of the Aldridge Commission report, NASA has a chance to make this change; a wrenching one for them since their whole history has had them in the forefront of exploration.

Such was the feeling at IBM in the 70s. They cared so little for their microcomputer division that they let the engineers in Boca Raton Florida give away the intellectual property for the hardware of their "PCs" to the world to copy. They felt so sure of their own market that they allowed Microsoft to write a contract that allowed Microsoft to sell a version of their DOS operating system to other companies without royalties. This revenue stream would one day rival IBM's total corporate income.

Transforming NASA and Enabling the Dream

NASA has the opportunity today to help change the world in a positive manner. They just have to realize that space is not their exclusive domain and that the astronauts that they hire with such fanfare will one day be only a part of the total number of people living and working in space.   The recently released report from the Aldridge Commission carries four interesting departures from previous studies, stretching back to the Collier's days. Recommendation 5-1 States:

> *The Commission recommends NASA aggressively use its contractual authority to reach broadly into the commercial and nonprofit communities to bring the best ideas, technologies, and management tools into the accomplishment of exploration goals.*

While past commissions have given nods to commercial activity, it has always been interpreted by NASA as a way to hand higher proportions of their contracts to Boeing, Lockheed, Northrop Grumman or other aerospace companies. This is a flaw that NASA could correct by following the path that IBM took in its transformation. The recommendation by the Aldridge Commission above, and the specific examples given below, carry the authority to do this, if NASA has the guts and the vision to take it.

Prizes (From the Commission Text)[iii]

> *Prizes. The Commission heard testimony from a variety of sources commenting on the value of prizes for the achievement of technology breakthroughs. Examples of the success of such an approach include the Orteig Prize, collected by Charles Lindbergh for his solo flight to Europe, and the current X-Prize for human suborbital flight. It is estimated that over $400 million has been invested in developing technology by the X-Prize competitors that will vie for a $10 million prize – a 40 to 1 payoff for technology.*

The Commission strongly supports the Centennial Challenge program recently established by NASA. This program provides up to $50 million in any given fiscal year for the payment of cash prizes for advancement of space or aeronautical technologies, with no single prize in excess of $10 million without the approval of the NASA Administrator. The focus of cash prizes should be on maturing the enabling technologies associated with the vision. NASA should expand its Centennial prize program to encourage entrepreneurs and risk-takers to undertake major space missions.

Given the complexity and challenges of the new vision, the Commission suggests that a more substantial prize might be appropriate to accelerate the development of enabling technologies. As an example of a particularly challenging prize concept, $100 million to $1 billion could be offered to the first organization to place humans on the Moon and sustain them for a fixed period before they return to Earth. The Commission suggests that more substantial prize programs be considered and, if found appropriate, NASA should work with the Congress to develop how the funding for such a prize would be provided.

The above recommendation is a very interesting introduction to what may well be the force that truly opens the door to the space frontier. The difficulty is that the Aldridge Commission has confused enabling technologies with spaceflight capabilities. The X-Prize, a $10 million prize that may well be won in the next few months, has not generated the $400 million in activity by focusing on enabling technologies. The teams have invested activity not to win $10 million, but because they want to develop the **capability for suborbital space flight**. The X prize would not have generated much interest if the goal were to develop a low cost engine, avionics package, simulation software for predicting ballistic flight paths, or a new astronaut glove. The commission did get this, but one gets the feeling that whoever wrote this section was trying to balance what NASA has already said they are going to do with prizes and what they knew in their heart would fire the imagination of the world!

Prizes have a rich history in promoting the growth of new technologies and concepts. As discussed in an earlier chapter, the first practical locomotive, appropriately named the "Rocket," was the result of a prize. The Longitude prize in the 18th century, offered by the British government, revolutionized navigation, saving thousands of lives, and led to the development of the first accurate maps of the planet to be drawn. In the 20th century, the Orteig prize inspired Charles Lindbergh to make his epic flight and help usher in the aviation age. Today, the X-Prize has been recognized by the commission as a major incentive for new approaches for suborbital spaceflight.

The last paragraph provides the solution. The idea for a multibillion-dollar prize for a manned landing on the Moon would literally transform our entire national psyche overnight as no other space activity since the Apollo era. We live in a different world today. We all know that, if you toss enough money at NASA, they could go back to the Moon. However, if we want to go and be able to really do this right, a prize would open the floodgates of innovation and entrepreneurship. The only thing that I would suggest is that it be for multiple billions and that they have to do it twice within a specified time period. This is the genius of the X-Prize. It is far easier to do something once than it is to do it twice. If you can do it twice, you can do it a hundred times. That is the difference between a stunt and a business.

The role for NASA in this new paradigm would be two-fold. One is their now traditional role of science. NASA is the appropriate agency to do amazing things, like flying spacecraft to Saturn, to Pluto and to the stars beyond. NASA is the appropriate place for scientist-astronauts, who will accompany the entrepreneurs and developers of a lunar settlement, mining and manufacturing enterprises, to study the terrain, examine the core samples for discovery and to advance the boundaries of human knowledge. The second role is in the IBM mold, as a developer of new technologies and capabilities that then would be licensed to the entrepreneurs, miners and developers of the lunar settlements, asteroid mining bases, and a Mars city. IBM reaps billions of dollars in revenues per year from their technology licenses, probably more than the entire current NASA budget.

These enabling technologies developed by NASA would be licensed to American companies with a fee paid back to the agency to plow into new developments in a virtuous cycle. It is only in the government or mega-corporations like IBM where long-term research and development can take place that is outside of the financial ability of small companies. This would help to break the negative cycle at NASA of cancelled programs and deferred dreams. Each NASA engineer would know that their work could help enable a new way to mine asteroids, power a spaceship, or provide a way to grow food on the Moon and Mars. There is a huge background of intellectual property at NASA now that would be of interest to commercial space companies. Robert Bigelow, a hotel chain magnate, is using NASA's development of an inflatable habitation module as the basis for his hotel in the sky. This hotel would be a destination for tourists and a home for businesses in space. NASA's developments of nuclear power sources, engines, software and other technologies would help enable many of the companies that would wish to compete for a lunar prize. NASA would still lead the development of some things including manned space systems, just like IBM builds computers, but the opening of the space frontier is going to take much more than just NASA and its contractors doing business as usual.

Tax Incentives

> *Tax Incentives. A time-honored way for government to encourage desired behavior is through the creation of incentives in the tax laws. In this case, an increase in private sector involvement in space can be stimulated through the provision of tax incentives to companies*

that desire to invest in space or space technology. As an example, the
tax law could be changed to make profits from space investment tax
free until they reach some pre-determined multiple (e.g., five times)
of the original amount of the investment. A historical precedent to
such an effort was the use of federal airmail subsidies to help create
a private airline industry before World War II. In a like manner,
corporate taxes could be credited or expenses deducted for the
creation of a private space transportation system, each tax incentive
keyed to a specific technical milestone. Creation of tax incentives can
potentially create large amounts of investment and hence, technical
progress, all at very little expense or risk to the government.

Tax incentives have a long and vital history in the growth of both the United States
and Britain in the Industrial Revolution. The National Railroad united the east and
west coasts of the nation just three years after the War Between the States, built with
a mixture of direct cash financial incentives, in the form of government guaranteed
bonds, and large grants of land. This provided a way for asset-based financing to be
gained by a group of shopkeepers in San Francisco and their counterparts in the east
and allowed them to change their 19th century world. In the 1840s and 50s,various
state governments, such as Virginia, actually purchased stock in railroad companies in
return for building railroads in their states. In the 20th century whole industries receive
tax breaks, loan guarantees and other financial incentives by states and the federal
government.

A few years ago, Congressman Dana Rohrabacher introduced into congress a bill
called "Zero G, Zero Tax," that would extend the same types of tax holidays that the
Internet enjoys into space. It would shelter the income of any company that actually
does something in space that makes money, excluding existing businesses such as
communications and remote sensing. The remote sensing industry (images from
space) has recently been bolstered by block buys of images by the U.S. military.
NASA could use similar incentives. One incentive would be to make the International
Space Station a free trade zone, free of government regulations and taxes. There is
much that can be done and it can be done the American way, by risk-takers, companies
and interest groups who see the promise that space has to provide a better future.

Regulatory Relief

Regulatory Relief. Government regulation of the nascent private sector space
industry is ongoing and will be necessary in the future, but it is important to ensure
that this industry not become over-regulated. A key issue in the private space flight
business is liability. There is a pressing need for a change in liability laws to set a
reasonable standard for implied consent. People throughout society do dangerous
things for fun and profit; it is not reasonable to impose governmental risk
standards on people who are willing and eager to undertake dangerous or

hazardous activities. In addition, numerous laws covering occupational safety and environmental concerns should be reviewed carefully to make sure that the government is not burdening new space industry unduly with irrelevant or unobtainable compliance requirements.

Regulatory relief is very necessary today to bring a measure of stability to the nascent suborbital tourism industry as well as other private enterprise launch efforts. Until recently, there were no regulations concerning the re-entry into United States territory of a reusable launch vehicle outside of NASA and the military. The Federal Aviation Administration (FAA) has struggled to deal with new forms of hybrid atmospheric/space vehicles and how to license them. Great progress has been made but much more needs to be done. For example, what are the tax implications and who is the taxing authority for a spaceship returning with a ton of PGMs to United States territory? Is it an import? What about licensing U.S. companies whose primary business is executed in space? The Congress will have to get involved to help in this arena, which, it seems, they may be willing to do.

Property Rights in Space

Property Rights in Space. The United States is signatory to many international treaties, some of which address aspects of property ownership in space. The most relevant treaty is the 1967 UN Treaty on the Peaceful Uses of Outer Space (the "Space Treaty"), which prohibits claims of national sovereignty on any extraterrestrial body. Additionally, the so-called "Moon Treaty" of 1979 prohibits any private ownership of the Moon or any parts of it. The United States is a signatory to the 1967 Space Treaty; it has not ratified the 1979 Moon Treaty, but at the same time, has not challenged its basic premises or assumptions.

The Commission is to be commended for addressing this issue. Property rights in space is a discussion whose time has come. There needs to be an appropriate way to recognize these rights by the United States government in such a way that will promote the exploitation of the Moon's resources. This extends beyond the Moon to the asteroids and Mars as well. Provision has to be made to recognize resource claims (water and nickel iron deposits) and provide legal protection so that entrepreneurs may obtain financing based upon legally established asset claims. This would go a long way to setting up the framework to allow us to move out into the solar system and develop its resource potential for the good of the people on the Earth.

How to Execute the Vision

In the next and final chapter of the book, I will lay out my proposed commercial architecture for the exploration and development of cislunar space and the Moon. However, to get to this point, I had to take you, the reader, through history, through the sources of our greatest problems today, and to the resources that will allow us to transcend the limits to growth. In presenting this architecture, I in no way claim that this is the only way to return to the Moon. In the past five decades, ideas about lunar architectures have evolved with our increase in knowledge about the resources of the Moon and the technologies that help to reduce cost. What I do hope to do is to stimulate thought on how this would be done if we were doing this with the goal of making money and establishing a settlement on the Moon. A commercial base has completely different drivers than an effort developed by NASA for the pursuit of science. These differences help to determine costs and to advocate certain decisions that do not make sense for a commercially driven plan.

We can create a commercial lunar base, settlement, and civilization in a manner that is environmentally responsible to the Earth and with the scale of resources that exist just in our planetary neighborhood. We can provide to the people of the entire earth a standard of living well beyond what a worldview based upon the limitations of our single planet could ever hope to achieve. However, before this glorious age appears we have to start from where we are today.

*   *   *   *   *

[i] W. MacDougall, The Heavens and the Earth, A Political History of the Space Age, Johns Hopkins University Press, 1985, P. 307

[ii] Ibid, P. 305

[iii] *Report of the President's Commission on Implementation of United States Space Exploration Policy* (The Aldridge Commission), June 2004, P. 34

Chapter 15:

An Architecture for a Commercial Cislunar Economy

As an engineer, when starting a design I first look at my market, the customers that want to buy my product, and how much they are willing to pay. That is basically what I did in the earlier chapters. I established, after a market study, that if I can sell PGMs at a discounted value, with a price set at about $296 dollars per ounce, then I have a very good chance at leapfrogging the competition, and more than that, it could help break open the market to grow more quickly. This is the breakthrough that the hydrogen economy needs. To lower costs to the point where fuel cells can compete with the internal combustion engine. This has been the secret to success in Silicon Valley for almost 30 years. This is also how, as far back as the beginning of the Industrial Revolution, the British Canals and the Steam engine helped that era begin its explosive growth.

How would one go about doing this for a commercial infrastructure to go to the Moon? What is the most inexpensive way of doing this that could have the quickest return on the dollar for potential investors? For this architecture I will ignore scientific exploration, as it brings nothing to the table in the beginning but requirements that drive up the cost. Later when the structure is in place science can then come along for the ride. Also, much of this work is based upon historical as well as recent NASA studies, especially the Orbital Aggregation & Space Infrastructure Systems (OASIS) group at NASA Langley.

Basic Assumption for a Cislunar Transportation Architecture

Use as much existing hardware as possible, to minimize development costs.

Launch Vehicles

This assumption leads to a divergence with many prior studies and the Aldridge Commission recommendation on this subject. Immediately, a heavy-lift rocket is eliminated due to the cost of development. A very conservative number for a heavy-lift rocket for the new exploration vision is $5-$10 billion dollars. For that price I can buy dozens of Delta Ivs, Atlas Vs, Ariane Vs, Proton's, Zenits, Japanese HIIs, and Soyuz vehicles. Besides the development cost there is another reason that I can eliminate any thought of a heavy-lift rocket, the International Space Station.

In the 1950s through to the 1980s the lack of a space station was a great impediment to the execution of space architecture due to its cost and the time required to build and outfit such a station. With the exception of Apollo, the architectures studied prior to the Stafford Report had a space station as a core required capability. In the early

Apollo days a million man-hours were dedicated to the question of Earth vs. lunar orbit rendezvous. The only reason that LOR was eventually chosen over EOR had solely to do with President Kennedy's deadline. Now that we have a space station in orbit over our heads, some experts at NASA, as well as the Aldridge commission, want to ignore its existence and again try and build a heavy-lift vehicle to promote an architecture that requires a major component that does not exist—and that everything else is dependent on.

There are legitimate reasons to have a heavy-lift vehicle that can lift 150 tons and this would be a wonderful thing if you could afford it. The problem is that you would not only have to pay for the development, but also the standing army that it takes to maintain the infrastructure that the system requires. With the bevy of medium heavy-lift rockets available today—listed above—heavy-lift is not absolutely necessary. A commercial return to the Moon can leverage these assets today and execute a block-buy—like Iridium did for their constellation of satellites—and get a darn good deal. Another example of this approach can be seen when EADS, a European defense contractor in Europe, purchased 30 Ariane V heavy launches for $3.6 billion dollars. Figure 15.1 illustrates a representative of the available medium heavy-lift launch vehicles available for commercial purchase today.

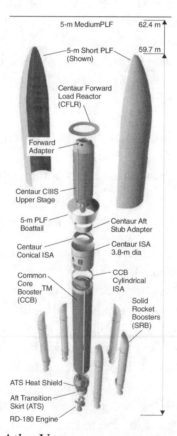

Figure 15.1: Atlas V (Picture Courtesy International Launch Systems)

There are six launch vehicles in production today that are in the medium heavy-lift category. The Atlas shown above is among the most reliable and cost effective launch vehicles on the market today. Table 15.1 gives a list of the vehicles and their performance to orbits of interest to us here.

Launch Vehicle	Payload 400 Km-51.6 Deg	Payload 35,756 Km GTO	Payload 100,000 Km	Payload to Trans-lunar Injection	Payload to Escape (C₃0)	Price Estimate $M
Atlas (552)[i]	18.1(23.4)	8.6(13.6)	7.6 (12.3)	7.5 (12.1)	6.5(11.0)	$150
Ariane V[ii]	20.2	10.05	8.0	7.7	6.8	$140
DeltaIV-H[iii]	24.1	12.7	11.1	10.5	9.306	$180
Proton[iv]	21.0(25.2)	6.12(12.1)	6.76(10.9)	5.7(10.5)	5.1(10.1)	$90
Soyuz[v]	5.9	2.15	1.95	1.7	1.6	$40
Zenit	6.5	6.00	5.3	4.8	4.0	$55

Table 15.1: Relative Performance of Different Commercial Launch Vehicles

All of the above numbers for the launch vehicles are estimates that vary based upon slightly different orbital parameters such as perigee height, inclination, fairing, and payload adapter used. These are close enough for our purposes here. The figures in parentheses are for currently planned upgrades that will enter the market within the next few years without any special funding. The pricing of these launch vehicles are fairly similar to each other with the exception of Soyuz and Zenit, with their lesser payload capacity, costing considerably less. I did not include prices for Chinese or Indian launch vehicles as they are for the most part lighter in payload and much more difficult to work with due to technology transfer considerations. I also did not mention the Japanese H-II because, although it is a very good launch vehicle, there is not a lot of data concerning its performance available. The numbers in parenthesis are for the Atlas V heavy, once speculative but now under active development, and for the Proton which will be soon adding a hydrogen/oxygen upper stage that will greatly improve performance.

These launch vehicles have similar payload fairings, internal payload volumes and launch environmental characteristics. Recently some of these companies have cooperated together and have standardized payload interfaces and documentation standards to decrease the burden on the commercial customer wishing to fly with them. One of the companies, Ariane Space, has an extensive accommodation capability for secondary payloads as well, some weighing up to one metric ton. With a wide variety of available systems it just makes sense to try and figure out how to use them to build a commercial cislunar architecture. To do this we need the second element in many of the historical architectures, a space station.

ISS and its Value to a Commercial Cislunar Architecture.

The ISS is an incredible asset for a commercial venture to utilize in lowering the cost for developing a cislunar infrastructure. The station, as it will be configured by 2010 is fairly close to the original Space Station Freedom design, excepting the addition of the Russian modules. Figure 15.2 shows ISS as it will be configured as of 2010:

Figure 15.2: ISS Assembly Complete Configuration (Picture Courtesy NASA)

While ISS does not have the dual keel design that was considered necessary (but not for the purposes of this book) for constructing and servicing interplanetary vehicles it does have many desirable features. ISS has two separate docking systems available, the Russian side where the Soyuz manned space vehicle and the Progress Space Freighter docks, and the American side where the Space Shuttle docks. Soon to be added to the fleet, are the European ATV, which will dock on the Russian side of the station, and the Japanese HTV, which will dock on the American side. Figures 15.3a and b show the Progress freighter and the manned Soyuz system:

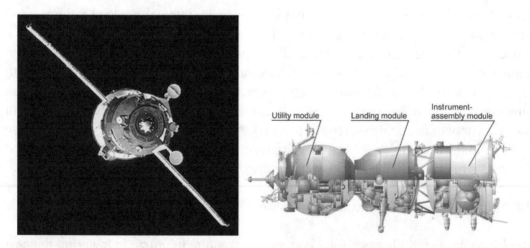

Figures 15.3 a, b: Russian Progress Freighter and Soyuz Human Transportation Systems
(Pictures Courtesy of ESA and RSA)

The Progress M1 space freighter, launched from Russia, can carry a maximum of 2230 kilograms of cargo. This cargo can be either equipment or supplies (up to 1800

Kg) or up to 1950 Kg of fuel to refuel the station keeping system on ISS.[vi] The balance is made up of dry or wet cargo. The progress is launched on a Russian Soyuz launch vehicle, one of the most reliable launch vehicles in the world. Over 100 Progress vehicles have been launched with zero failures. The Progress used for ISS is a slightly improved version of the system that was used to support the Russian MIR space station. The Progress is an unmanned system. The aperture of the hatch between the Progress and the space station is 80 centimeters in diameter. The Progress is also available for sale today to commercial customers.

The Soyuz shown above in figure 15b carries up to three people to ISS and is launched from Russia on the Soyuz launch vehicle. The Soyuz spacecraft and vehicle both have been in production for over 30 years with an unmatched safety record. Only one crew has ever been lost in the Soyuz and that was on the first flight in 1971. The new Soyuz TMA, now in production, can accommodate larger crew persons, up to 190 cm (6'2.5" tall), has a new digital computer, better parachute, and many other improvements for safety and operations.[vii] Due to the toxic nature of the fuel in the Soyuz propulsion system every six months a new vehicle has to be carried up to ISS. This is a "ferry" mission and has been used in the recent past to carry paying space tourist-explorers to the station. The Soyuz is also available today for purchase by paying customers. The Russian Space Agency and Energia (The Russian equivalent to Boeing) have entered into such deals in the past and are quite willing to do so in the future. The Russian state space program out of necessity has become quite entrepreneurial in selling both the Progress and Soyuz seats and vehicles since the fall of the Soviet Union.

Supplementing the Progress for unmanned cargo missions are two new vehicles that will enter service by the end of 2005. The first of these is the European ATV, which will fly to ISS for the first time in late 2004. The second of these is the Japanese HTV, which will make its inaugural flight sometime before the end of 2006. Figure 15.4a and b show these two vehicles:

Figure 15.4a, b: European ATV in Flight and Japanese HTV Docked at ISS
(Pictures Courtesy ESA and JAXA)

The European ATV is on the left and the Japanese HTV is on the right in the above picture. Despite the difference in appearance the two vehicles have exactly the same purpose, the delivery of cargo to ISS. Table 15.2 gives the relative operating characteristics of each compared to the Progress.

Vehicle	Launch Vehicle	Size (Meters)	Dry (M Tons)	Wet (M Tons)
ATV[viii]	Ariane V	4.5 X 10.1	5.5	4.4, 0.86, 0.84
HTV[ix]	H-II	4.4 X 9.2	7.0 (pressurized)	N/A
Progress[x]	Soyuz	6.6 m³	1.8 (2.23 total)	1.95

Table 15.2: Cargo Capacity of the Three Available Cargo Carriers

The availability of three different cargo carriers gives an impressive capability and flexibility to deliver cargo to the station by three different launch vehicles. The ATV and the HTV have a high volume dry cargo compartment that can store large amounts of supplies. The ATV docks autonomously to the Russian side of ISS via a system similar to the KURS that is used by the Progress vehicle. The HTV does not dock on its own, but flies up to the station and is captured by the Space Station Remote Manipulator System (SSRMS) and is docked to the Common Berthing Mechanism (CBM), located on the nodes on the American side of the station. The HTV has the ability to carry up complete ISS internal racks and is the functional Equivalent of the Multipurpose Logistics Module (MPLM) carried up on the United States Space Shuttle. I am not going to discuss the cargo carrying capabilities of the Space Shuttle, except in passing, as it is expected to be phased out by the year 2011. Also, until that time the Shuttle will be occupied with carrying up cargo needed to complete the ISS and will not be available for commercial cargo.

With the robust capabilities of the three cargo vessels and the availability of the Soyuz manned spacecraft, a commercial operation can plan, with confidence, operations between the ground and the station for high value cargo. For large modules that need a dedicated launch the launch vehicles indicated in table 15.1 suffice. The large dedicated cargos carried up would have their own navigation system and would rendezvous with ISS and be captured in a similar manner to the Japanese HTV. In order to minimize development costs the Proton would be the standard cargo carrier as it has already been qualified in this area. Some extra work may be necessary for the Delta IV and Atlas V. Large payloads could be captured and delivered to ISS via a commercial space tug under development by CSI. Now that we have established that payload carriers exist and can be used to ferry large amounts of cargo to ISS next we turn to the capabilities at ISS for assembly, test, and servicing of vehicles and payloads destined for cislunar space.

ISS Capability for On Orbit Assembly

The International Space Station has a good robotics capability to support the assembly of space vehicles bound for cislunar space. Figure 15.5 illustrates the Crew and Equipment Translation Aide (CETA) cart:

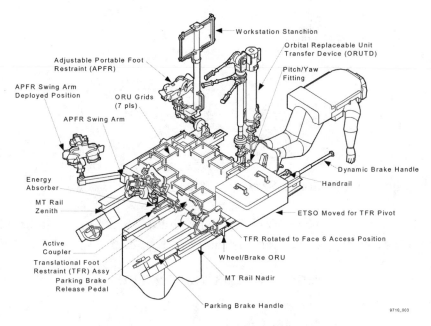

Figure 15.5: Crew Equipment Translation Aide (CETA) Cart (Illustration Courtesy NASA)

The CETA is used to move the crew, EVA equipment and cargo along the entire length of the space station truss. It has many attachments that can be of use to crew persons doing various jobs, including assembly tasks during ISS construction. Currently there is one of these carts at ISS with another one due to be delivered within the next two years. Two crew persons at a time can work on the cart or one can be attached to one cart that carries the Space Station Remote Manipulator System (SSRMS) and one can be on the CETA cart. This, along with one potential addition may be all that is necessary in terms of infrastructure to allow for the assembly of cislunar bound spacecraft.

Figure 15.6 shows the SSRMS attached to a Mobile Transporter (MT) that also slides along the rails of the ISS truss. The Mobile Transporter and the SSRMS robotic system, coupled with the Special Purpose Dexterous Manipulator, makes up the ISS Mobile Servicing System (MSS). This allows the transportation of large payloads along the ISS truss, such as large parts of a lunar bound spacecraft. This system can be controlled from inside of ISS at the robotic workstation. Therefore the need for EVA by human crewpersons is somewhat minimized by using this system. The SSRMS can even be used to capture large payloads rendezvousing with ISS.

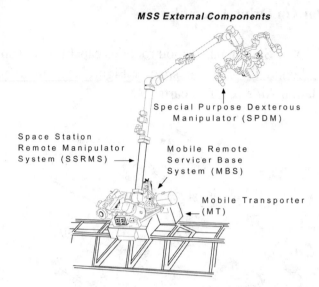

Figure 15.6: Mobile Servicing System (Illustration Courtesy NASA)

There may be one further piece of robotic hardware that would need to be added to this already very capable system. This is shown in figure 15.7:

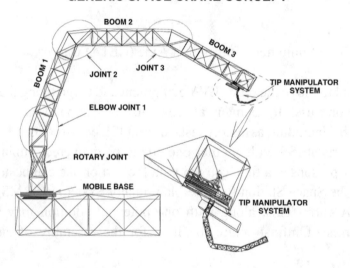

Figure 15.7: NASA Langley Space Crane (Illustration Courtesy NASA LaRC)

The Langley Space Crane (LSC) is designed to move or hold very large, very heavy payloads such as vehicles being assembled. The LSC mounts to the Mobile Transporter already located on the station or fixed directly to the space station main truss. This is not a new development other than for final space qualification. NASA Langley built versions of this crane and have tested it at the center in their robotics lab as well as in the neutral buoyancy tank (water tank) at the Marshall Space Flight Center (MSFC) in Huntsville Alabama. Elements of the space crane have been flown in space and assembled. The LSC forms the basis of a lot of our commercial work as variations of this crane would be used on the Moon as well. Plate XIII shows these cranes in action during the construction of the lunar cargo lander at ISS. Before we delve into the assembly process I want to lay out the other elements of the architecture.

Beyond LEO

In the 1970s and 80s during the development of architectures to return to the Moon, whether or not a space station was involved, heavy-lift launch vehicles were always needed. It was just assumed to be the case because that is what we did with Apollo and the development of the Saturn V. However, this is not a requirement based upon physics. That is, other ways of doing this are possible and in order to deliver a low cost total systems solution the comparative costs need to be addressed. NASA has done some very interesting studies related to such architectures and has come up with a very interesting solution. I want to present their solution and then see how it can be modified with existing assets to lower costs.

OASIS

Pat Troutman from NASA's Langley Research Center (LaRC) led a lunar mission architecture study called the Orbital Aggregation & Space Infrastructure System (OASIS) whereby they looked at non heavy-lift options for a cislunar architecture.[xi] This architecture did not demand the existence of a 150-ton heavy-lift launch capability. Their idea began with this premise:

> *Minimize point designs of elements in support of specific space mission objectives and maximize modularity, reusability and commonality of elements across many missions, enterprises and organizations.*

What this means basically is that a "point design," is a code word for a heavy-lift launch vehicle. The study did assume that a small upgrade to the Delta IV-H would increase the lift mass to 35 metric tons with a larger 6 meter shroud. This was to reduce costs but I will modify their architecture somewhat to mitigate this need as I will show. Their ideas for maximizing modularity, reusability, and commonality of elements is the strength of their proposed architecture and is very similar to my approach. Again, I should point out that much of the work to develop these architectures originated at NASA and that I am merely tinkering with the design approach to further minimize cost while ignoring institutional impediments that NASA has to deal with concerning foreign purchase of hardware and services. The system elements shown here are integrated into a complete architecture for the return to the Moon.

OASIS Architecture Overview

To me the OASIS architecture is a very good starting point for a commercial cislunar infrastructure. It has most of the right elements for that. It ignores, for the most part, heavy-lift, (it does assume an upgrade to the Delta IV to 35 metric tons to ISS and a 6 meter fairing which I do not), and it utilizes a way-station located at the Earth/Moon L1 libration point. For those who do not know, the L1 libration point is a quasi-stable gravitational point between the Earth and the Moon that is at the "top" of the gravitational "hill" between the Earth and Moon. Objects on the Earth side of the hill "fall" into an orbit around the Earth and objects on the lunar side of the hill fall into an orbit around the Moon. There are five of these librations points in cislunar space but we are only concerned with L1 in this book. Figure 15.8 shows the NASA OASIS architecture:

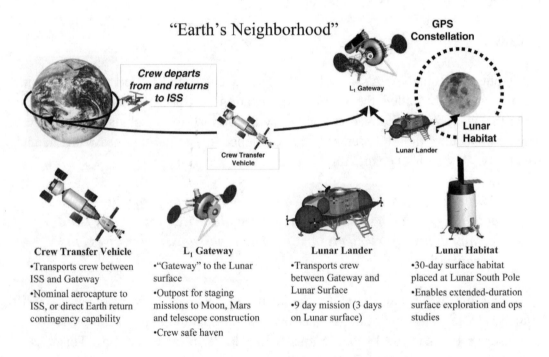

Figure 15.8: OASIS Cislunar Transportation Architecture (Picture Courtesy NASA)

Central to the success of this architecture is the existence of the L1 "Gateway" outpost. This outpost has habitation capability for a human crew. The outpost also has multiple docking ports to support the docking of "Hybrid Propulsion Modules," or HPMs. These modules are basically fuel tanks to provide fuel for the lunar lander that is based at the Gateway outpost. What the existence of the Gateway outpost does is lessen the logistics burden on the cislunar transportation system. No longer does a lunar lander have to piggyback a ride with an orbiter like in Apollo, or ride on the top of a heavy-lift launcher that would otherwise be needed.

This is called forward-basing of supplies and fuel. With refueling at the Gateway outpost an option, now your Crew Transfer Vehicle (CTV) only has to go between the Earth and the Gateway outpost. Also, the lunar lander does not have to return to the Earth and can become a reusable system that ferries between the Gateway outpost and the lunar surface. These factors liberate this architecture from dependence on heavy-lift launches. All of this calls for the establishment of the L1 Gateway outpost, as an inexpensive space station stopover point where supplies can be stored and fuel made available for incoming vehicles.

In this architecture a CTM is a vehicle that is launched to ISS empty on an EELV expendable launch vehicle. This CTM then mates with HPM and a Crew Transfer Vehicle (CTV) that can move the HPM full of fuel and the CTM to the L1 Gateway outpost. The HPM is then used to fuel the Lunar Landing Vehicle (LLV), which was previously brought to the Gateway outpost via a Solar Electric Propulsion (SEP) space tug. All of this activity takes place at ISS, a station requiring very little modifying to support this use. The crew then transfers from the CTV to the Gateway outpost for a few days of provisioning and fueling of the LLV. After this is accomplished they board the LLV and take a three day journey to the Moon, where they land, spend three days on the surface, and then return to the Gateway outpost. The crew then reverses the process, transfers back to the CTV. The CTV, now mated to just the CTM (the HPM is left at the Gateway outpost), returns to ISS.

This is a very attractive architecture and all of the components of the system are fully reusable. All that has to be provided to repeat the process is more fuel. In the next few subsections I will describe in detail the components of this alphabet soup and then I will slightly rearrange it and marry it to the architecture that I propose that goes beyond this exploration and science oriented architecture. However, it must be noted that this was a NASA study and that there are more voices at NASA than just the ones advocating large heavy-lift vehicles. NASA headquarters is or recently has been considering a similar architecture as one of four options moving forward with the return to the Moon.[xii]

Hybrid Propulsion Module

The Hybrid Propulsion Module (HPM) is a flying fuel tank, carrying liquid oxygen and hydrogen for use in transferring payloads between the earth and an L1 outpost. It also is used to carry the fuel to an L1 outpost which can then be loaded onto a lunar lander destined to land on the Moon. Figure 15.9 illustrates the HPM concept:

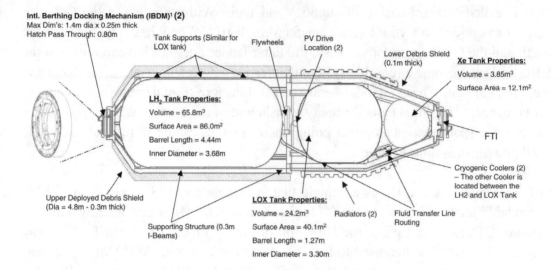

Figure 15.9: Hybrid Propulsion Module (HPM) (Picture Courtesy NASA LaRC)

The HPM would be a core component of this stepping stone approach to reaching the Moon. With a dry mass of ~4000 kilograms the HPM could be launched to the orbit of ISS with as much as 20 (Delta IV-H) or 21 (uprated Proton) metric tons of propellant. The total capacity of the HPM is 31.2 tons of propellant. The principal propellant carried is liquid hydrogen and oxygen. A secondary Xenon tank is carried as well (refer to the figure), that carries enough Xenon for the SEP tug to return to ISS from the L1 Gateway outpost. The most technologically advanced aspect of the HPM is that this is a "Zero Boil Off" system. What this means is that none of the liquid propellants would get warm and boil away during storage. This feature overcomes the principal objection to using extremely cold liquid propellants in the past. This is why the HPM is a very expensive fuel tank as well.

According to a study conducted by Boeing for NASA, the development cost for the HPM would be about $611 million dollars with no reserve funds added. The cost for production of the first unit would be about $130-$170M dollars with some savings for units after that. Most of this cost comees in the development of the insulation and propellant management systems for the HPM. To date, no one has ever built a long-term zero boil-off tank system for space and this would be a major and necessary advance in technology that would be well worth the cost.

In any architecture studied you would have several of these, at least four in operation for a robust propellant storage and transfer system. The liquid propellant stored in the HPM would be adequate to fuel either a CTM or an LLV (Cargo Transfer Vehicle and Lunar Landing Vehicle in case you have forgotten). This would power a trip between ISS and the L1 Gateway outpost (CTV) or between the Gateway outpost and the lunar surface and back to L1. It would take only one large and one light launch from the ground to fully fuel the HPM which would cost less than $250 million—worst case. So if you had to rely on propellant launched from the Earth the fuel and recurring transportation cost would be about $500M dollars. Now let's look at the CTV and its characteristics.

Chemical Transfer Module (CTM)

The CTM is actually a pretty slick looking cargo transfer system. Figure 15.10 shows its interior and its functions.

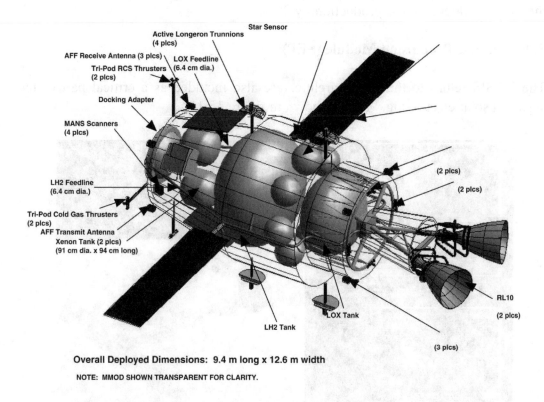

Active Longeron Trunnions (4 plcs)
Star Sensor
AFF Receive Antenna (3 plcs)
LOX Feedline (6.4 cm dia.)
Tri-Pod RCS Thrusters (2 plcs)
Docking Adapter
MANS Scanners (4 plcs)
(2 plcs)
(2 plcs)
LH2 Feedline (6.4 cm dia.)
Tri-Pod Cold Gas Thrusters (2 plcs)
AFF Transmit Antenna
Xenon Tank (2 plcs) (91 cm dia. x 94 cm long)
RL10
(2 plcs)
LOX Tank
LH2 Tank
(3 plcs)
Overall Deployed Dimensions: 9.4 m long x 12.6 m width
NOTE: MMOD SHOWN TRANSPARENT FOR CLARITY.

Figure 15.10: Chemical Transfer Module (CTM) (Graphic Courtesy NASA)

The CTM is designed as a heavy duty fast payload transfer system. The configuration shown here has two Pratt & Whitney RL-10 engines for over 44,000 pounds of thrust. It can interface to and take fuel from an HPM module. The CTM also has the ability to dock with a Crew Transfer Vehicle after dropping off the HPM at L1 and returning it to ISS for full reuse. This would be NASA's first heavy-duty fully reusable chemical space tug. And for commercial use it makes a lot of sense. It relies on mostly existing technology and it is designed for up to 200 missions, which should do a lot to amortize the cost and would be the most reused space vehicle ever built. The dry weight of this CTM was ~4200 kg, easily lofted by any of the existing launch vehicles. It could also be launched with most of its first trip's fuel. The rest made up by another tanker at our now growing tank farm at ISS.

The cost of the CTM as estimated by Boeing is reasonably priced. The estimated development costs were about $1.5 billion dollars using a 30% reserve fund for contingencies. The first unit cost would be ~$180 million—quite reasonable considering that it will be used 200 times. This level of reuse would result in a per-flight cost of $1.8 million or $6.5 million counting development costs, but not counting operations and fuel. Further flight copies would be lower in cost. There is every chance that the development costs would be as much as $400 million lower as Boeing put in the cost of the development of a new hydrogen/oxygen engine, which is probably not needed with the existence of the Pratt & Whitney RL-10. Certainly a commercial mission would not undertake this development, thereby lowering cost of the total development and production by 20%.

Solar Electric Propulsion Module (SEP)

The OASIS return to the Moon architecture also included as a critical part of the system a solar electric tug. This is shown in figure 15.11

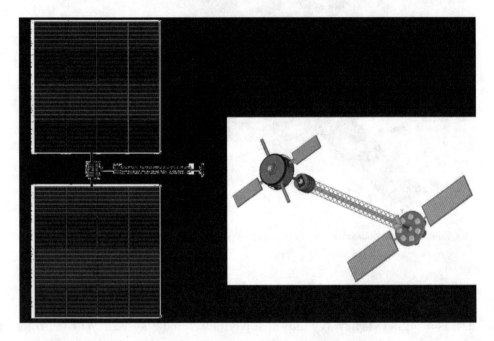

Figure 15:11 Solar Electric Propulsion Module (Picture Courtesy NASA)

The purpose of the SEP is to provide a highly efficient low thrust space tug traveling between LEO and an L1 Gateway outpost or beyond. This space tug could be used to move the lunar lander from LEO to L1. It could also transfer other heavy payloads such as the L1 station itself out to that location.

The SEP would be powered by 2700 square meters of solar array, generating enough power to run nine, 50-kilowatt gridded ion thrusters. These thrusters would cumulatively consume 450 kilowatts of power and would produce two Newtons of thrust in total. With this SEP the transit time for a 10,700 Kg lunar lander would be about 6 months. The cost to develop the SEP would be high. The Boeing study estimated that the development costs would be $1.88 billion. Also, the unit cost for the first item was expected to be almost $500 million. That is a lot of money for a system that only gets used twice a year due to the length of time it takes to do the transit.

Even if this had a lifetime of a hundred round trips it still means that the round run cost is $5 million and the loaded cost $23 million versus the $6.5 million for the CTM. The savings comes in the form of fuel efficiency. It takes about 3700 kilos of Xenon to take a heavy payload like an LLV from LEO to L1. Being in 28.5 degrees inclination and 51.6 means little to this mission as the plane-change to the inclination of L1 is done at this altitude with little penalty. The 3700 kilo propellant usage means that a much cheaper Soyuz or Zenit launcher could be used as well as an Atlas III or equivalent launcher. It has to be noted that this would require a separate HPM (that has not been costed) which could be as simple as a pressurized tank. If it were brought up in an HPM on an Atlas Heavy or a Proton, it would make economic sense to do so.

However, since the 3.7 metric tons of Xenon is only 12% of the total payload weight the cost for fuel per flight is ~$10.75 million per round trip. Add this to the $23 million to pay back the vehicle cost for a total of $33.75 million per flight and the value is there. This compares to the 1.2 Proton flights required to fill up the tanks of the CTM, or about $106.75 million per flight for fuel plus $6.5 million for amortization of costs for a total of $113 million.

One hundred trips for the SEP is an unrealistic number due to the radiation damage inflicted on the system in its passage through the radiation belts. An optimistic number with today's technology is 20 round trips. The number of round trips decline the cost increases rapidly per trip. For 20 trips the SEP amortization costs increases from $23 to $118 million added to the fuel cost of $10.75 million, totaling $128.75 million, closing the gap with the chemical system.

The truth is that until we have flown both of these systems a while we will not know what their lifetimes are. However, the above shows that there is a relationship between total development costs versus run costs. Clearly, the run costs are lower with the ion system but the development costs erase the benefit.

NASA SEP Alternative

This is where I begin to diverge with the NASA OASIS study. There was a systems design done for a proposal to NASA to rescue the Hubble Space Telescope that went to the approximate level of detail as the Boeing study here. Figure 15.12 shows this system with the Hubble Space Telescope attached.

Figure 15.12: HST Attached to the HTARV

We estimated that for an 88 kilowatt SEP based upon existing Hall Effect Thrusters (or HETs, a different form of ion propulsion) rather than gridded ion thrusters, and using existing communications satellite solar arrays run in parallel, that the total development costs would drop to about $200 million. This number also assumes that the SEP is assembled at ISS from components rather than launched as one big package. The total costs for the first item from the assembly line would be about $350 million if you increase the solar array power to 100 kilowatts.

The four Hall Effect Thrusters used for this mission would generate over 7 Newtons of thrust versus the 2 Newtons of thrust for the gridded ion thrusters. The downside is that the exit velocity of the Xenon is about 60% (Isp of 1875 for Hall and 3,300 for gridded) for a HET relative to a gridded ion thruster. This translates into almost doubling, to six thousand kilograms of Xenon per round trip to L1. However, the development cost is $200 million versus $1.8 billion for the gridded ion system. This is because we are using existing thrusters (HT-220HT from Pratt & Whitney), and existing solar array designs from high power communications satellites. This system would also use existing computers and software from Swedish Space Corporation that is already proven for boosting an HET powered spacecraft from a lower Earth orbit to the Moon.

With a total cost of $550 million versus $2.3 billion the per flight cost drops to 25% of that of the NASA gridded SEP design, making the 20 round-trip system economically competitive with the CTM on a per-mission cost basis. The primary payload for the SEP is the lunar lander and HPM fuel modules and it is unclear whether chemical or electric propulsion is a clear winner. What that means is that if costs are really driving the system the SEP could be deferred until later in the program unless really heavy payloads beyond the capability of the CTM are needed, which may not be the case.

Crew Transfer Vehicle (CTV)

The CTV, used to move crews to and from ISS would be fully reusable. Figures 15.13 and 15.14 show the overall outline of the system.

5.5m

Figure 15:13: Crew Transfer Vehicle Outside and **Figure 15:14:** Inside
(Picture Courtesy NASA)

This CTV would be attached to the HPM/CTM for the trip from ISS to the L1 Gateway outpost. This version of the CTM would carry three people but the OASIS group also looked at a four seat version. The nice thing about this design is that it is optimized for the job that it does. This lowers the development costs versus something like the current NASA thinking for the Crew Exploration Vehicle or Constellation as it is called now.

In the Boeing study the CTV development costs were estimated at ~$1.2 billion and the production costs at $236 million. These actually seem to be somewhat reasonable for the aerospace world today. Since the CTV is fully reusable the costs can be amortized over a lot of missions, just like the CTM. If it were to be reused 200 times the cost per mission would be an inexpensive $7.8 million dollars.

Commercial CTV Alternative

As good as those numbers are a commercial entity would want to do better if possible. That possibility exists in the form of the Russian Soyuz. There is absolutely no reason that the Soyuz could not be used for this mission. One variant of the Soyuz was originally going to be the Russian equivalent of the Apollo Command and Service

Module. Soyuz has demonstrated in Earth orbit the ability to operate as an independent vehicle for up to 17 days with 2 crew, and can carry three people in a fair amount of comfort for a shorter duration. Remember that Soyuz has been in production and flying for 37 years, a lesson in production economy to all of us. A commercial company, Constellation Services International has proposed to NASA to fly the Soyuz on a cislunar flyby trajectory that could just as easily go to the L1 Gateway Station. Figure 15.15 illustrates the CSI concept.

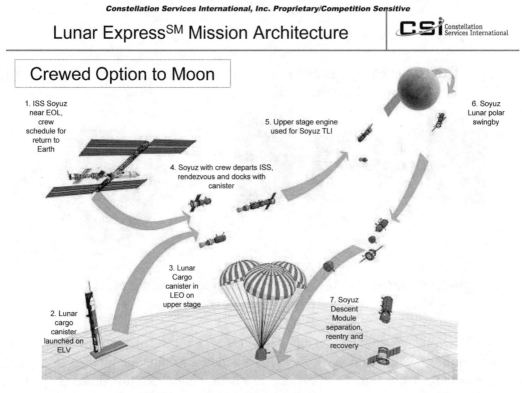

Figure 15.15: CSI©'s Lunar Polar Flyby Mission (Picture Courtesy CSI)

The key element of this architecture is the launch, after the Soyuz has docked to the International Space Station (ISS), of a cargo module which remains in orbit attached to an upper stage. This cargo module would contain the all the additional subsystems required to extend the capability of the Soyuz during its cislunar mission, such as life support for the crew and high-gain antenna for communications with Earth. The Soyuz would depart from ISS, dock with the cargo module/upper stage, and then the upper stage engine would be used to send the entire stack on a free-return lunar swingby trajectory or on a trajectory to L1 as desired.

CSI© Chief Technical Officer Ben Muniz related to me that this would require an upgrade to the heat shield of the Soyuz but that this kind of heat shield was flown by the Russians on the Zond series of spacecraft that flew circumlunar missions in the late 1960's. The Zond and Soyuz return modules are quite similar; the Zond was actually designated as model "Soyuz 7K-L1" and was a modification of the original Soyuz 7K-OK model used for early Earth orbit missions. Therefore, the heat shield change would not significantly modify the current Soyuz design (although specific design impacts must be assessed). The Soyuz TMA is heavier than the CTV (7,220 Kg versus 5282 for the CTV), but this is still well within the range of payload weight that the CTM/HPM combination could transfer to the L1 Gateway outpost. Initial research shows that the new heat shield could increase this weight by as much as 300 kg, although that remains to be verified.

This opens up the interesting possibility that, for a commercial mission, the cost of the CTV would be deferred (not eliminated). The Soyuz option is a throw away one that over time would not be cost-effective relative to the CTV but it would allow for an early capability and it would allow the more effective use of the development money. Instead of trying to fund all of these system's developments at once you could use the Soyuz for about ten missions or at least 3-6 years, and stage the development of the CTV until after the other parts of the architecture are in place and operating. The use of the Soyuz does reduce the crew by one over the favored NASA model from OASIS but this can be to our advantage.

Reusable Lunar Landing Vehicle (LLV)

The OASIS team at NASA Langley married their cislunar architecture to a JSC related proposal for a single stage reusable LLV. The ideas are complimentary to each other as they both use the L1 Gateway outpost the forward base for resupply of fuel, provisions, and crew change out. Figure 15.16 illustrates the JSC lander idea:

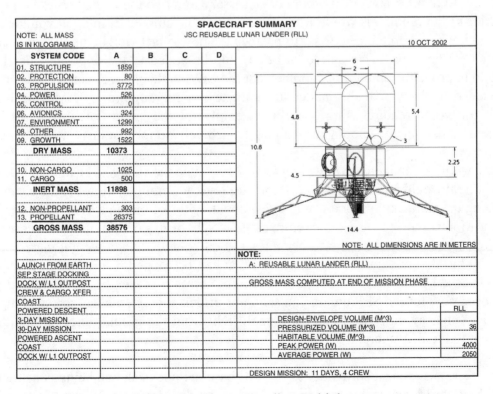

Figure 15.16: JSC Reusable Lunar Landing Vehicle (Diagram Courtesy of NASA)

JSC did a very good job on this design for a reusable lander. There is only one issue with it that bears examination. Currently it is sized for four people and the diameter of the habitable portion is six meters. This is sized for a possible upgraded Delta IV-H that would itself cost a lot of money to develop. If we back off slightly, and if we are using the three man Soyuz, we should be able to reduce the volume to the twenty five square meters that a five meter diameter habitable volume would produce. Figure 15.17a, b shows the internal volume of the lander and the vehicle sitting on the lunar surface.

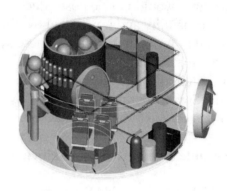

Figure 15.17a, b: LLV on the Surface and Internal Layout (Renderings Courtesy NASA)

In order to make this work it would be necessary to move the lander's airlock to the outside of the cabin, this would free up a considerable amount of internal volume and would improve habitability. Also, to balance the now offset weight, an internal cargo carrier of the same shape as the airlock would be added to the opposite side. This would improve the amount of supplies that the LLV could carry. A small crane could be added to the top of the lander to facilitate moving supplies off. Making these changes would require using on orbit assembly for the lander but this is far less expensive than the development of a new launch vehicle for just this one payload.

There would be an increase in weight to accommodate the changes. I estimate that the dry weight would increase to about 12,500 kg for dry weight to allow for the changes. Adding the extra cargo space would allow about 1,500 kg more payload to be carried to the Moon or an increase in the return cargo weight of 1,000 kg. With these increases the dry weight fully provisioned becomes 14,000 kg and the wet weight increases to 31,200 kg, the maximum propellant load that can be carried in the HPM. The total mass of the LLV is increased to 45,200 kg. The increase in weight does tend to drive the need now for the SEP unless you send the airlock and external cargo carrier up to the L1 Gateway outpost via an additional flight from the ground directly, and then mate it to the LLV there.

In Plate 13 a similar lunar lander to the JSC baseline one is shown being constructed at ISS. Plate 14 shows this lander in its landing on the Moon, while plate 15 shows a one way cargo version being disassembled for its useful parts.

The cost of the JSC derived LLV was not given in the study material that I researched. However, even if the cost was in the $2 billion range to get the first model it would be a bargain as it would be fully reusable. If we used existing hydrogen/oxygen engines such as the RL-10, and use on-orbit assembly to eliminate the requirement to build a new launch vehicle there is no reason that the LLV could not be designed, built and flown to L1 for about $3 billion. The second off the production line would cost less than $500 million. You would want two in order to have a rescue ability for the first crew in place at the L1 Gateway outpost.

L1 Outpost

The last component in this cislunar architecture before we discuss the base itself is the L1 Gateway outpost. Neither the OASIS or the JSC reports that I have access to discuss the L1 Gateway in much detail. One thing that is clear about the NASA version is that it is built on the model of the now defunct Transhab inflatable module developed by JSC. It is my idea, based again upon minimizing development, to use an ISS Node for the basis of the L1 Gateway outpost. It has life support capabilities that are already being used on ISS as well as several docking ports, vitally necessary for a waystation like this outpost.

Figure 15.18 shows my concept of the L1 outpost, combined with a SEP for power and station keeping:

Figure 15.18: L1 Gateway Outpost (Picture Courtesy SkyCorp Incorporated)

This outpost is built of either existing systems such as the ISS node or ones that will already be developed for this architecture such as the SEP. The SEP that carries the L1 Gateway outpost up to its L1 home remains with it, to provide power, station-keeping, and excess power to run a water electrolysis unit. This water electrolysis unit would be used to generate hydrogen and oxygen from lunar water, if available, or to process water brought up to the L1 Gateway outpost via direct flights from the Earth. The advantage of sending water to the L1 Gateway outpost and processing it there is that the tanks for water would be far cheaper than HPMs and you only need a limited amount of them for this architecture. A Russian docking adapter would have to be added, but this could be done at ISS.

Actually the entire L1 station would initially be attached to ISS for outfitting and testing before the SEP tug and the Gateway station separate and head on its way to L1. This allows full up testing of the Gateway outpost in LEO where any problems can be identified and fixed before it gets deployed to its final orbital location. The Node and any ancillary hardware needed would be launched via the now plentiful Delta IV-H's and Atlas Vs and maybe a Proton if necessary. This is the power of on-orbit assembly to provide this type of construction and testing services for a robust capability in L1.

Revised Architecture for Cislunar Space Access and Architecture Summary

In figure 15.8 the OASIS architecture for cislunar space access is laid out. Figure 15.19 illustrates my modifications to that architecture.

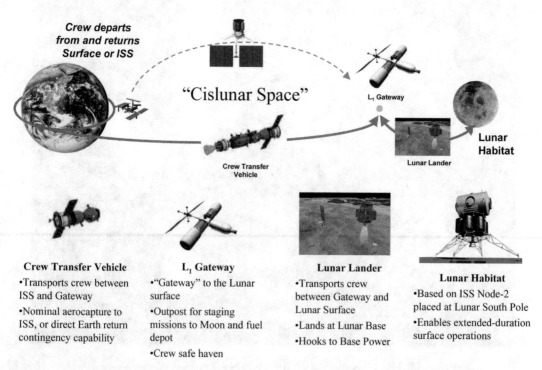

Figure 15.19: Modified Cislunar Architecture for Commercial Return to the Moon

What is not shown in the illustration above are direct flights from the Earth to the Moon to emplace payloads that are light enough not to require the L1 Gateway outpost services. To me the most attractive part about this architecture is the utilization of as much existing hardware as possible. This allows the designers to focus on the really important parts of the mission and gives them fixed specifications to work with, in terms of launch vehicle lift weight, fairings, and ISS accommodation capabilities.

Costs for this architecture can be developed as well through the first lunar landing. Table 15.3 outlines these costs and the execution sequence:

Sequence Event	Launch Vehicle	Activity	Cost
ISS Modifications	European ATV	Modify ISS	$600 million
L1 Node and SEP	DeltaIV/AtlasV	Assemble outpost	$1.2 billion
L1 Outfitting	Progress	Outfit outpost	$100 million
L1 Transit	—	Move to L1	$10 million
SEP/HPM	Atlas V/Delta IV	Assemble SEP	$750 million
LLV Launch 1	Atlas V/Proton	Assemble at ISS	$700 million
LLV Launch 2	Proton	Outfit and some fuel	$250 million
HPM Fueling	Proton	Fuel for HPM	$271 million
CTM	Atlas V/Delta V	Transport for outfit	$350 million
Soyuz	Soyuz	Humans to ISS	$55 million
CTM/HPM/Soyuz	CTM	Humans to L1	$2 million
LLV to the Moon	LLV	Humans to Moon	$50 million
Total Cost to First Landing			**$4.3 billion**

Table 15.3: Cost to First Lunar Landing (Not including Development)

This is a very reasonable cost based mostly on the work by Boeing for costing the elements of the OASIS architecture. This has to be followed by the expected development costs for the elements mentioned in this chapter.

HPM Development	$795 million
CTM Development	$1.5 billion ($1.1 w RL-10)
SEP Development	$1.8 billion ($250 million)
CTV Development	$1.3 billion
Total	$5.7 billion ($2.2 billion)

By eliminating the CTV for the first missions and using existing hardware for the SEP and CTM, the cost for returning to the Moon can be as low as $6.5 billion dollars for a three day mission similar to Apollo 17. This mission would be able to land at the lunar poles, something that was never possible in the original Apollo architecture.

We Want More

Going to the Moon for three days is clearly inadequate for anything serious that we want to do. This sequence was shown to illustrate how to do a focused recon of a base site by humans after robotic missions have already landed. The three day mission could set up homing beacons, landing aids, and scout out the base area for the best location for the next flurry of automated landings that would begin building a base. This is what we turn to next in our final chapter.

*   *   *   *   *

[i] Atlas Launch System Mission Planners Guide, International Launch Services, September 2001
[ii] Ariane 5 Users Manual, Issue 3, Revision 0, Ariane Space Customer Service, March 2000
[iii] Delta IV Payload Planners Guide, MDC 00H0043, The Boeing Company, April, 2002
[iv] Proton Launch System Mission Planner's Guide, Revision 5, International Launch Services, December 2001
[v] Soyuz User's Manual, ST-GTD-SUM-01 Issue 3 Revision 0, The Soyuz Company, April 2001
[vi] http://www.astronautix.com/craft/proessm1.htm
[vii] http://www.astronautix.com/craft/soytmasa.htm
[viii] http://www.estec.esa.nl/spaceflight/atv.htm
[ix] http://www.nasda.go.jp/projects/rockets/htv/component_e.html#specification
[x] http://www.russianspaceweb.com/progress.html
[xi] spacecraft.ssl.umd.edu/ old_site/design_lib/OASISEXEC_97.pdf
[xii] www.lpi.usra.edu/lunar_return/connolly.pdf

Chapter 16 :

Initial Lunar Base and Final Thoughts

In the following lunar base design, I will describe how to do the initial set up and operation. There is no way that anyone can adequately forecast beyond that point how such a settlement will evolve. The key is that it *is* a settlement, something more than a temporary habitat for three people for a few days. The base developed here will have as its initial primary focus, the identification and extraction of resources. Science will be of a lesser priority. It will be a profit center derived by serving the needs of the scientists who come here but not exclusively designed for their use. This is the difference philosophically between what we have had in the recent past from NASA and what I propose. Just as before in the cislunar architecture, success is measured by cost control. Not skimping on testing or safety, but selecting intelligent systems engineering solutions that lower cost and improve the ability to generate a return on investment.

Lunar Base Site

In the 1960s what was speculation is now strongly established theory awaiting a return to the Moon to prove or disprove. In the last chapter I wrote of an initial lunar scouting mission that would basically do a site survey for the final location of the lunar base. The principal driving criterion of the lunar base comes down to one very large requirement: Sunlight.

The Earth rotates at an angle of 23.45 degrees away from "north" in our solar system. This is what causes our seasons on the Earth. In the summer in the northern hemisphere the sun shines for months at a time as the North Pole points toward the sun. In the winter this is reversed. The Moon's rotational angle is only 1.5 degrees, therefore its "seasons" are far less pronounced than on the Earth. This opens up the possibility that there are locations in the polar regions where the sun shines all year. On the Earth, the low angle to the horizon attenuates solar energy in the polar regions. This is due to the longer path through the atmosphere that the rays have to travel to get to the ground. The Moon does not have this problem. Therefore, if there were any locations at the lunar north or south poles where there was permanent sunlight in summer and winter, a lunar base would be far less expensive to set up. Nuclear power, frequently added to lunar architectures, would not be needed at a polar base, at least not for a good while. This saves a considerable amount of money for our commercial development efforts.

In the early 1990's Clementine, the Defense Department's technology test satellite, mapped the polar regions of the Moon multiple times over a period of several months. Since the Moon rotates on its axis at the same rate that it goes around the Earth, a "day" on the Moon is approximately 28 Earth days long. Clementine was able to image these polar areas over several lunar "days", allowing a map of the distribution of light and shadow in the polar regions to be developed. Scientists at the Lunar and Planetary Institute and others have taken these maps of the light and dark areas to develop a darkness/illumination map. It is unfortunate that Clementine did not operate for a full year in lunar orbit so that these darkness/illumination maps would represent both "winter" and "summer" views but nonetheless these maps indicate that there are areas at the lunar poles where the sun shines for most, if not all of the time. Figure 16.1 is a preliminary darkness/illumination map from the Lunar and Planetary Institute of the lunar poles.[i]

Lunar Polar Composites
Cumulative images of the pole throughout one lunar day

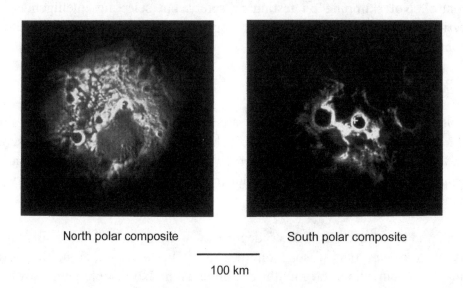

North polar composite   South polar composite

100 km

Figure 16.1: Darkness/Illumination Maps of the Lunar Polar Areas (Image Courtesy LPI)

These maps reveal to scientists not only where the sunlight shines most of the time but also where it is dark most, if not all, of the time. Note that these maps were generated during the northern summer and southern winter, accounting for the different distributions of light and dark.

In a recent paper at the Lunar and Planetary Institute's yearly conference Dr. Ben Bussey and three other scientists gave a paper on this subject.[ii] The scientists conclusions were that there are no areas in the south polar regions where the sun shines all of the time. However there are areas that are close together, (next to the cross mark in the right hand image) where the sun shines close to 90 percent of the time. This considerably relaxes any requirements for energy storage as the lunar night is reduced from 354 hours to 212 hours at several locations. The three highest points together reduce the hours of lunar night down to about 78 hours in the south. In the north, during the summer the paper's results were that there were four locations between the two small craters at the 10:00 and 12:00 locations on the large crater in the north where the sun shines 100% of the time! However, it is not known whether these four areas are illuminated in like manner in the winter. Be that as it may, this finding has great potential to lower the cost of any potential lunar base.

The really nice thing about the darkness/illumination maps is that they also point to the locations where the sun shines zero percent of the time, indicating areas where water or other hydrogen bearing molecules are located in concentrations of interest to a lunar base. Refer back to plates 9 and 10 for a mosaic of pictures of the lunar polar areas and plate 7 showing the areas where water or hydrogen bearing molecules (hydrocarbons) are located. It just so happens that the areas where the sun shines most of the time are located very close to the areas where the sun shines the least amount of the time! Since the scientific community's estimates vary from ten million to one billion tons of water, it is clear that if these resources exist, the cost of fuel can be cut considerably by producing both the oxygen and hydrogen locally. While this is all good news it is also speculative until proven so, therefore I will initially ignore this potential, and will not have local water production as part of the cost equation. It just seems to make sense, that in order to control costs by eliminating nuclear power and access to potential local water, that we place the base in the permanently sunlit areas, probably at the North Polar location.

Lunar Base Logistics and Hardware Sequencing

The weakness of the cislunar architecture presented in the previous chapter is that you can't put a lot of "stuff" on the lunar surface at one time unless you use on-orbit assembly at ISS or at the L1 Gateway outpost. However, with the lift capability that we have today, and the capability of modern systems this does not seem to be as much of an issue as it would have been in the past. While scientific exploration of the Moon demands that missions be spread out over large areas, economic considerations, as we have seen drive us to the most cost effective locations at the poles. The logistical considerations that confront the lunar base designer after *where* has been decided is *what* and *how*. The lunar base site at the poles has eliminated one "what", that is nuclear power, so it seems that the first "what," is power. With the first "what" decided, the engineers look at available solar power systems to see what is available. They do a preliminary design, and then see how to get it to the Moon, keeping in mind the logistical constraints of existing launch vehicles and the logistical infrastructure at ISS and the L1 Gateway outpost.

Initial Power

Engineers know that at the Earth's orbital distance from the sun that the density of solar energy is about 1358 watts per square meter. The engineers look in the catalog and see that today the best solar cells have a guaranteed efficiency of 28.5% and are fairly radiation tolerant. Cells are available from two companies, Emcore and Spectrolab, so competition is not a problem. Taking the cells from either company and designing them into a solar array using the best techniques today results in an array with a power density of about 325 watts per kilogram.

Taking out our catalog of launch vehicles—and after asking our propulsion engineers what kind of performance can be had from hydrogen and oxygen lander engines—we see how much payload we can land. This turns out to be about 4000 kilograms, if we use a Proton or Atlas V-heavy for a direct flight. Or closer to 10,000 kilograms if you go through the L1 gateway with a cargo version of the manned lander. However, if you have an array with the power density above and use 37.5% of the payload capacity for power processing, and some power cables, and other ancillary hardware, you get a power density per lander of 81.25 kilowatts for a direct flight and over 200 kilowatts for a heavy flight.

Since the electrical engineers say that 81 kilowatts is much more manageable than 200 kilowatts we now delve a little further. Today power conversion technology has incredible efficiencies over what was available in the SEI days. Back then power conversion efficiency averaged 89%. Today, on a regular basis, it is as high as 97%. Using a slightly more conservative number of 95% and considering losses in the power cables, this gives us 75 kilowatts of power availability from one launch. Looking at power demand for a minimal base, 75 kilowatts is plenty. Figure 16.2 shows our power lander descending to the lunar surface toward a sunlit landing site:

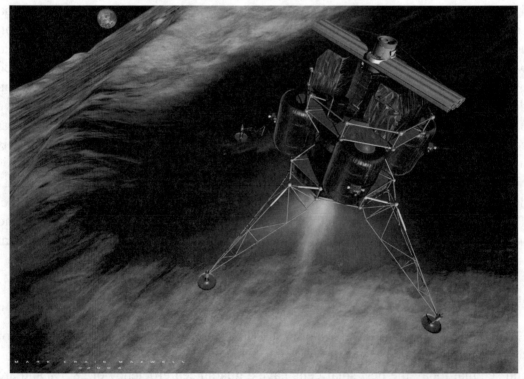

Figure 16.2: Power Lander On the Way to the Lunar Surface

Plate 13 shows two of these landers on the surface and one dropping down to join them. Before the next human crew was sent to the Moon at least two of these power landers would land and deploy their arrays. For backup battery power fuel cells would be used.

The cost for a power lander would be primarily driven by the solar array cost. Using a conservative number of $1000 per watt for a completed array gives a cost of $81 million for the array. The lander itself is driven by the propulsion system costs. I estimate this to be about $40 million dollars. With structure, avionics and the rest costing about $10 million, the total cost of a power lander is approximately $ 131 million. With the cost of a Proton at $90 million and operations at about $5 million the total cost to emplace one power lander is ~$226 million dollars. This cost will decline after the first unit but a conservative number for two of these would be about $450 million to put 150 kilowatts of power on the Moon.

Habitation

After power, a place to initially live seems to be the next logical step in the process. Taking another page from our "don't develop it if it can be purchased" book, an initial habitation module would be derived from what we have already used for the core of the L1 Gateway station, an ISS node module. This is shown in figure 16.3:

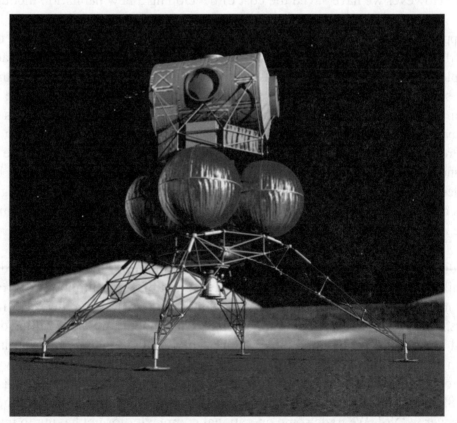

Figure 16.3: ISS Node 2 Type Core Habitation Module

As may be obvious from the picture, it is highly unlikely that this cargo lander and ISS Node could be launched on a single launch. This would be assembled on orbit at ISS from a Node that we lifted to ISS via a Delta IVH or Atlas V. The components for the lander would have also been taken up on one of these vehicles. The mated lander and Node would weigh about 22,000 kilos dry mass. The tanks would get one HPM full of fuel as the cargo lander is a one-way affair, and would be ferried to L1 via the SEP electric tug. This might take too long and so another option might be to use a CTV and HPM to go ahead of the LLV heavy and hab and they could meet back up at the L1 Gateway station. After a final check-out and fueling, the LLV heavy and hab would make their way to the lunar surface and land near to the power landers.

The cost of this mission would be considerable. It takes two launches from the Earth to carry up the Node and the components for the LLV heavy. That is $220 million on a Delta IVH or Atlas V again. By this time I better be getting a darn good discount on launches which is figured into this price. I have been told by various people that if you were buying Nodes from a production line and outfitting them with life support equipment similar to that on ISS, that it would cost around $500 million dollars. Since this would be the second one purchased it stands to reason that the costs should be somewhere near that number. The power lander is a slightly upgraded version of the standard LLV lander with a stronger support structure for the increased weight. With an estimated cost of $500 million for the LLV this brings the combination up to a $1 billion. However we have saved the cost of developing a new habitation module.

To support the flight we need an Atlas V with an HPM, costing $181 million dollars plus $120 million for the launch. We also need a CTV requiring fuel—another example where a Proton filled with water could be used, launched to ISS and the water separated into fuel from the ISS arrays and/or the SEP arrays while at the station. As an aside, when the SEP is at ISS it can produce about 50 kilowatts of on-orbit average power for the station which could be sold to NASA or other customers. The Proton filled with water would cost about $100 million, including the water container ($10 million). This water could also be provided by low-cost launch vehicles such as the SpaceX Falcon V. While it has limited capacity, providing about 4 tons of water, it could make up the difference for the partially full HPM launched on the Atlas V or Proton. This would cost about $12 million per mission.

So adding to the $1 billion for the hab, LLV, $220 million for their launch, and about $20 million for assembly related tasks we have a cost of $1.24 billion for that portion of the mission. Add up all of the related tasks here and we have another $413 million dollars. Finally, we have a total cost to emplace a hab module at $1.637 billion dollars. Would heavy lift have lowered the cost? The total cost for launch this way is $452 million dollars. This would have taken at the very least one Shuttle C class heavy lifter and probably two, as volume is more of a constraint than weight for these missions. Each Shuttle C is going to cost $500 million dollars, not including development, so the answer is 'No'. We trade some operational complexity for not having to maintain the heavy-lift launch vehicle and the inherent problems that entails.

This mission scenario takes about 8 months to unfold due to the slow pace of the SEP in climbing the gravity hill between the Earth and L1. During this time other direct flights to the Moon take place.

Infrastructure

Since we do not want the hab module to sit 35 feet in the air on the top of the lander it stands to reason that we need a crane and other hardware to get the hab removed from the lander and put in place. We also need other support hardware for when the crew arrives. I estimate that this will require two more direct-flight lander missions. The first lander will carry the components of an overhead crane that will be used to remove the hab module from the cargo lander. This is shown in figure 16.4:

Figure 16.4: Langley Overhead Space Crane for the Moon

Even though this crane is large the components are lightweight and would all fit on one direct flight lander. The crane has its own drive system and would be able to lift the Node from the lander and take it to wherever it is decided to place it. A very preliminary estimate for the cost of this crane is no more than $30 million. Mate that with a, now somewhat lower cost, lander (down to $200 million as production picks up), and an Atlas V at $120 million, and you have a total cost of around $350 million to put the very useful crane in place.

Next on the list are infrastructure support items like those seen illustrated in figure 16.4. There will be various carts, a rover for the humans to move around on, and our little friend George 101. Now George is a curious fellow. His father was named Robonaut and was developed by NASA and DARPA. What makes him George is that I mate him with Dean Kaman's Segway scooter technology. The Segway's ultrastable platform makes this a natural marriage for the Moon. George can be operated via telepresence, or operate in autonomous mode to carry out certain limited tasks. Multiple Georges can work together to carry out cooperative tasks, such as build the overhead crane. Figure 16.5 is George 101:

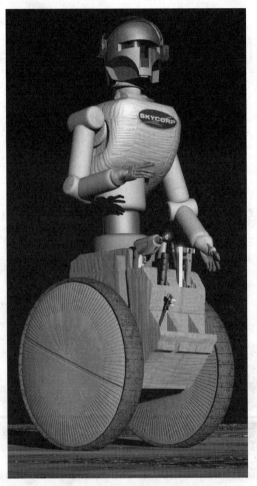

Figure 16.5: George 101

George is probably the most sophisticated robot ever built for lunar development. His propulsion system is derived from the Segway as mentioned earlier. His imaging system has multiple wavelengths, from infrared to ultraviolet. His dexterous "hands" can use tools or can be changed out and replaced directly with tools. George will be very strong, with the ability to lift several hundred pounds on the Moon. George can also pull carts of materials around on the Moon and two Georges working together could unload cargo from landers, disassemble or assemble hardware, and generally be of tremendous use to the crewpersons on the Moon.

George will not be cheap, costing about $100-$200 million to develop. However, he should be fairly inexpensive to build, with an estimate in the $10 million range. George is fairly light, weighing in at about 250 kilograms on the Earth. On the first mission a total of four Georges would be stacked on a standard lander, along with long lengths of electrical cable, fuel cell batteries, carts and other hardware needed to pull the whole initial base together.

ISRU

There is one other component to the system that has to be included from day one. This is In Situ Resource Utilization (ISRU). Without ISRU this whole exercise is futile. To some people, the Moon is poor in resources, to be written off as a slag heap to pass on the way to Mars. Nothing could be further from the truth. I went through some of the resources in earlier chapters, including PGMs. However, the first product produced at the base will be oxygen, iron, magnesium, and possibly silicon, and aluminum. This all comes direct from the native regolith and rocks. Some nickel/iron meteorite material is included in this, up to 1% as determined from Apollo samples.

It is unfortunate that over the last 30 years NASA has not put much money into ISRU research. We do know how to extract oxygen and various metals from the regolith and rocks but not on an industrial scale. We know what processes to extract oxygen and metals are most efficient, but we really do not know which ones are the most cost effective. To say that we have definitive ways of making ISRU work on an industrial scale today would be a stretch of the imagination. However, we do know what we have to do in order to learn the answers to these questions. In this regard the ISRU that I first propose be used at a lunar base for oxygen and iron extraction is simple heating in an induction furnace.

In the technical language of the scientist's in this field, this is called Vapor Phase Pyrolysis. There has been some very high quality work in this field. One proponent who did a lot of work at NASA JPL was Wolfgang Steurer. He wrote a paper for the Space Studies Institute's conference on lunar manufacturing in 1985 where he gave the results of his calculations and experiments in this area.[iii] Steurer's thesis is that the easiest way to get oxygen from the lunar regolith is by simple heating in a closed container. This may not be as efficient as some chemical methods but it is very simple and he was able to obtain good results from his experiments in this area. Figure 16.6 is my idea of what this would look like on the Moon:

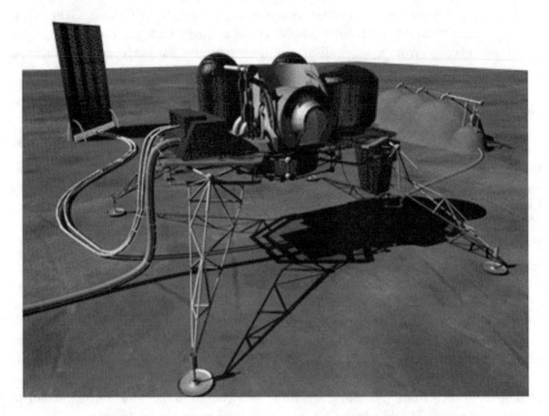

Figure 16.6: Test Vapor Phase Pyrolysis Unit and Smelter

Steurer determined by computational means, that by heating regolith to a temperature of between 2500-3000° Kelvin (2226-2726° Centigrade/4040-4940° F) that you could extract about 19.6% of the oxygen by weight from the regolith. This is without any further processing. He selected these temperatures to be easily achievable with solar energy or by electrical heating. Steurer used a slightly more conservative number of 17% as a general case without any regolith processing. My initial thought in this area is to use a mixture of solar and electrical heating.

Steurer also talked in his paper about metals extraction using this process. His idea was for the use of temperature-controlled plates that would preferentially extract certain metals. He was able to successfully extract silicon via this method. As of the date of this particular paper he was able to verify, by experiment, oxygen extraction at a similar percentage value as his calculations.

In this paper Steurer had some good estimates for power requirements, the weight of the production facilities, and the yield per year. His first idea was to produce 100 tons of oxygen per year from this facility. In our scenario, this would provide the oxygen load for four HPMs per year. This facility would require the input of about 658 kilowatts of solar power. The energy required would be provided by a mirror several meters in diameter, or this could be provided by 258 kilowatts of electrical power. In our scenario, this works out to four power landers total with a fifth needed for general power consumption. The weight of the installation was estimated at 12,100 kilograms which could be provided with one heavy cargo LLV.

The cost of this system would not be high. In the metals industry on the Earth there is extensive experience with vacuum induction furnaces and one of these designs could be adapted for our use. The power supplies would basically be off-the-shelf items modified for conductive and radiative cooling. I estimate that this would cost no more than about $100 million to develop, another $200 million for the lander, and $450 million for transportation costs to the Moon. This would put in place on the Moon a capability to cut by 5/6ths the logistics requirements for fuel for HPMs. By this time there are a lot of these hanging out at the L1 Gateway station empty. If they could be filled with LUNOX then we could dramatically drive down our transportation costs. If water were found and easily extractable (not a sure thing by any means), then the costs would be very inexpensive compared with terrestrial launch.

The Assembled Base

With all of the elements mentioned in this chapter so far we have all of the things needed for a lunar base. I would add at least two more direct flights of landers with provisions to support the humans and spares to support the Georges and other hardware. The total cost for those missions would be about $500 million dollars as we can benefit from consistent production runs of hardware and better pricing. The humans return to the Moon after all of the above is in place and the Georges have assembled the overhead crane and deployed the power system cables to the emplaced hab module. This flight will be by three people for a six month stay.

Their first task is that of setting up the base, hooking up the power systems, assembling everything and moving the supplies sent from the earth to the base. The Overhead crane will bring the landers, intentionally landed a couple of hundred meters from each other, together. At this time space is pretty tight and some of the crew sleeps in the lander and some in the Node. As the ISRU unit is set up, more power landers arrive, a total of five, to provide power for the ISRU unit. This, along with the rest of the base, is set up and running by month three. Plates 3 and 4 show activities associated with base setup. After three months of operation by the end of the sixth month, the tanks on the ISRU unit are filled with oxygen. These tanks were removed from the Hab's cargo lander, as well as the rest of the control electronics hardware. This provides enough fuel to get the human LLV to get back to L1, although at this time this is not needed since the LLV landed with enough fuel to return. The LLV would be fully fueled anyway with the oxygen in order to take that back to the L1 Gateway station as a way to cut costs on the next trip. Figure 16.7 shows the first product from the iron derived from the oxygen production:

Figure 16.7: First Lunar Building Made with In Situ Materials

The iron here came as a byproduct of oxygen production. The induction furnace has a dual role to play to melt the iron derived from the oxygen production so that it can be formed into plates and beams, the simplest structural elements. The Quonset hut would be used as an equipment storage facility for the Georges and even for landers. The roof here would be high enough for the crane to drive into. Welding would be simple on the Moon as vacuum-welded joints are very strong and tend to have low defects. This would especially be the case with the Georges doing the welding.

This Quonset hut would be set up directly behind the Node hab used by the crew. This would probably be done by the second crew but an interior wall could be built using iron plate and sealed. A transit tube between the Node and the hut would then be constructed and the first large habital space on the Moon opened up. The base would begin to build up from here and the ground truth surveys of nickel/iron deposits and lunar water would begin. Table 16.1 gives a cost breakdown on how much it would cost to emplace a base such as the one described here, summarized from this chapter.

It is probably obvious by now for those who live and breath this at NASA that these numbers are considerably lower than a lot of existing estimates. However, many of the numbers that I use here are directly from NASA studies. Even if I remove some of my lower numbers and go with the high ones from NASA, this would still cost no more than $20 billion dollars.

Base Element	Development Costs ($M)	Production Costs ($M)	Launch/ops Costs ($M)	Number Used	Total Costs
Power Lander	$200	$131	$95	5	$1330
Hab	$500	$500	$637	1	$1637
Crane	$30	$230	$120	1	$380
George	$200	$10	$120	1 (4 units)	$330
ISRU	$100	$300	$450	1	$850
HPM	$611	$181	$120	6	$1697
CTM	$1500	$180	$120	2	$2100
CTV	$1200	$236	$120	0	$0
SEP	$200	$350	$120	3	$1140
L1 Outpost	$500	$850	$240	1	$1590
Soyuz	$0	$30	$25	3	$165
Cargo	$0	$200	90	2	$580
LLV	$2000	$1000 (500)	$450	2	$3950
Extra Fuel	$10	$10	$90	4	$410
Totals	$7051	N/A	N/A	N/A	**$16,159**

Table 16.1: Total Lunar Architecture Costs to First ISRU Production

Final Thoughts

A question that I asked a few chapters ago was, "Can a lunar base be built at a reasonable cost." I think that the answer to this is yes and that this base, oriented toward "In Situ Resource Utilization" applications would rapidly drop in total costs as compared with the above numbers that did not get past the third mission. However, even with this third mission, ISRU begins to pay for itself as the extra oxygen in the LLV transits back to the L1 Gateway outpost.

After the third mission of course comes the fourth. During this time an additional $1.1 billion is spent sending five more power modules to the Moon, bringing the total power output to 600 kilowatts. Add to this another ISRU unit, this one tailored to melting metals while the original would be dedicated to oxygen. If there is water then the ISRU unit would be used to bake it out of the regolith and capture it. As the inventory of metals start to increase they are used to make a landing pad, roads, and structures that can hold atmosphere to prepare for larger crews.

After the fourth crew additional crews of six people would be sent, two crews at a time. It may be that we never build the CTV on the Earth but on the Moon. All of the cargo landers are stripped of their tanks and engines except one which is reconfigured as a hopper to carry crews on nickel/iron ground-truth prospecting. The engines and new tanks made on the Moon are used to build a large CTV possibly using mostly lunar derived materials, to transport crews as large as six from ISS back to the Moon.

It is highly speculative to talk about the cost for PGM mining at this state but these resources would be found early on and begin to be exploited first for the nickel and iron locally and then for the PGMs to send to the Earth. This could easily happen for an additional $2 billion to get the first kg returned. So for probably $20 billion you get to that point. This is no more than what the big dig in Boston cost and it cost less than many other infrastructure projects. This is the crux of the whole matter. We are told that we need the money to solve our Earthly problems when the facts are that we need to spend this money going to the Moon so that we can address our problems down here.

If the government is too wrapped up in their own issues to be able to do this, it is not outside the bounds of private enterprise. Deals of this size are done all the time, and think what having access to and rights over a billion kilos of platinum would do for your corporate portfolio.

There are many reasons to go and no convincing ones not to. If the limits to growth on the Earth are real, and if the Earth really is in the balance, then it just make sense to use the Moon's resources to tip the balance in our favor.

The only limits that we have are in our imagination.

*   *   *   *   *

[i] http://www.lpi.usra.edu/publications/slidesets/clem2nd/slide_32.html

[ii] D. Bussey, et al, *Permanent Sunlight at the Lunar North Pole*, Lunar and Planetary Science XXXV, March 2004, Houston TX, P. 21387.pdf

[iii] W. Steurer, *Lunar Oxygen Production By Vapor Phase Pyrolysis*, Space Studies Institute Conference on Lunar Manufacturing, 1985 TL 795.7.S653

Apogee Books Space Series

#	Title	ISBN	Bonus	US$	UK£	CN$	
1	Apollo 8	1-896522-66-1	CDROM	$18.95	£13.95	$25.95	_____
2	Apollo 9	1-896522-51-3	CDROM	$16.95	£12.95	$22.95	_____
3	Friendship 7	1-896522-60-2	CDROM	$18.95	£13.95	$25.95	_____
4	Apollo 10	1-896522-52-1	CDROM	$18.95	£13.95	$25.95	_____
5	Apollo 11 Vol 1	1-896522-53-X	CDROM	$18.95	£13.95	$25.95	_____
6	Apollo 11 Vol 2	1-896522-49-1	CDROM	$15.95	£10.95	$20.95	_____
7	Apollo 12	1-896522-54-8	CDROM	$18.95	£13.95	$25.95	_____
8	Gemini 6	1-896522-61-0	CDROM	$18.95	£13.95	$25.95	_____
9	Apollo 13	1-896522-55-6	CDROM	$18.95	£13.95	$25.95	_____
10	Mars	1-896522-62-9	CDROM	$23.95	£18.95	$31.95	_____
11	Apollo 7	1-896522-64-5	CDROM	$18.95	£13.95	$25.95	_____
12	High Frontier	1-896522-67-X	CDROM	$21.95	£17.95	$28.95	_____
13	X-15	1-896522-65-3	CDROM	$23.95	£18.95	$31.95	_____
14	Apollo 14	1-896522-56-4	CDROM	$18.95	£15.95	$25.95	_____
15	Freedom 7	1-896522-80-7	CDROM	$18.95	£15.95	$25.95	_____
16	Space Shuttle STS 1-5	1-896522-69-6	CDROM	$23.95	£18.95	$31.95	_____
17	Rocket Corp. Energia	1-896522-81-5		$21.95	£16.95	$28.95	_____
18	Apollo 15 - Vol 1	1-896522-57-2	CDROM	$19.95	£15.95	$27.95	_____
19	Arrows To The Moon	1-896522-83-1		$21.95	£17.95	$28.95	_____
20	The Unbroken Chain	1-896522-84-X	CDROM	$29.95	£24.95	$39.95	_____
21	Gemini 7	1-896522-80-7	CDROM	$19.95	£15.95	$26.95	_____
22	Apollo 11 Vol 3	1-896522-85-8	DVD*	$27.95	£19.95	$37.95	_____
23	Apollo 16 Vol 1	1-896522-58-0	CDROM	$19.95	£15.95	$27.95	_____
24	Creating Space	1-896522-86-6		$30.95	£24.95	$39.95	_____
25	Women Astronauts	1-896522-87-4	CDROM	$23.95	£18.95	$31.95	_____
26	On To Mars	1-896522-90-4	CDROM	$21.95	£16.95	$29.95	_____
27	Conquest of Space	1-896522-92-0		$23.95	£19.95	$32.95	_____
28	Lost Spacecraft	1-896522-88-2		$30.95	£24.95	$39.95	_____
29	Apollo 17 Vol 1	1-896522-59-9	CDROM	$19.95	£15.95	$27.95	_____
30	Virtual Apollo	1-896522-94-7		$19.95	£14.95	$26.95	_____
31	Apollo EECOM	1-896522-96-3		$29.95	£23.95	$37.95	_____
32	Visions of Future Space	1-896522-93-9	CDROM	$27.95	£21.95	$35.95	_____
33	Space Trivia	1-896522-98-X		$19.95	£14.95	$26.95	_____
34	Interstellar Spacecraft	1-896522-99-8		$24.95	£18.95	$30.95	_____
35	Dyna-Soar	1-896522-95-5	DVD*	$32.95	£23.95	$42.95	_____
36	The Rocket Team	1-894959-00-0	DVD*	$34.95	£24.95	$44.95	_____
37	Sigma 7	1-894959-01-9	CDROM	$19.95	£15.95	$27.95	_____
38	Women Of Space	1-894959-03-5	CDROM	$22.95	£17.95	$30.95	_____
39	Columbia Accident Rpt	1-894959-06-X	CDROM	$25.95	£19.95	$33.95	_____
40	Gemini 12	1-894959-04-3	CDROM	$19.95	£15.95	$27.95	_____
41	The Simple Universe	1-894959-11-6		$21.95	£16.95	$29.95	_____
42	New Moon Rising	1-894959-12-4	DVD*	$TBA	£TBA	$TBA	_____
43	Moonrush	1-894959-10-8		$TBA	£TBA	$TBA	_____
44	Mars Volume 2	1-894959-05-1	DVD*	$28.95	£20.95	$38.95	_____
45	Rocket Science	1-894959-09-4		$TBA	£TBA	$TBA	_____
46	How NASA Learned	1-894959-07-8		$25.95	£18.95	$35.95	_____
47	Virtual LM	1-894959-14-0	CDROM	$TBA	£TBA	$TBA	_____
48	Deep Space	1-894959-15-9	DVD*	$TBA	£TBA	$TBA	_____

CG Publishing Inc home of **Apogee Books**
P.O Box 62034 Burlington, Ontario L7R 4K2, Canada
TEL. 1 905 637 5737   FAX 1 905 637 2631
e-mail marketing@cgpublishing.com
* NTSC Region 0

Many more to come! Check our website for new titles.
www.apogeebooks.com